Internet of Things and Unmanned Aerial Vehicles-based Applications for High-Yield Precision Agriculture

The text focuses on the basic algorithms used for hard problems or multi-objective problems in precision agriculture. It comprehensively discusses precision agriculture techniques such as the use of global positioning systems, crop sensors, geographic information systems, remote sensing, and soil sampling.

Key Features:

- Presents the Internet of Things–based methods for the detection of diseases and water requirements in precision agriculture.
- Focuses on the basic algorithms used for hard problems or multi-objective problems in precision agriculture.
- Covers applications of unmanned aerial vehicles and machine learning in precision agriculture water management.
- Explains geospatial technologies for the management of pests and diseases in crops.
- Showcases case studies on leveraging the Internet of Things in agriculture for high yields.

It is primarily written for senior undergraduates, graduate students, and academic researchers in the fields of electrical engineering, electrical and communications engineering, computer science and engineering, agricultural science and engineering, and information technology.

Communication Networks and Smart Techniques for 5G and Beyond
Edited by: Nisha Gupta

Secure Communication for the 6G and the Internet of Things Networks
Advances and Applications
Edited by Gaganpreet Kaur, A.V. Senthil Kumar, G. Prasanna Lakshmi, and Rakesh Nyayapati

Internet of Things and Unmanned Aerial Vehicles-based Applications for High-Yield Precision Agriculture
Edited by Amit Bage, Prakash Pareek, Aman Kumar, Dharmendra Prasad Mahato, and Rardchawadee Silapunt

Internet of Things and Unmanned Aerial Vehicles-based Applications for High-Yield Precision Agriculture

Edited by
Amit Bage, Prakash Pareek, Aman Kumar,
Dharmendra Prasad Mahato, and
Rardchawadee Silapunt

CRC Press
Taylor & Francis Group
Boca Raton London New York

CRC Press is an imprint of the
Taylor & Francis Group, an **informa** business

Designed cover image:

First edition published 2027
by CRC Press
2385 NW Executive Center Drive, Suite 320, Boca Raton FL 33431

and by CRC Press
4 Park Square, Milton Park, Abingdon, Oxon, OX14 4RN

CRC Press is an imprint of Taylor & Francis Group, LLC

ISBN: 9781032964966 (hbk)
ISBN: 9781032964980 (pbk)
ISBN: 9781003589730 (ebk)

DOI: 10.1201/9781003589730

Typeset in Sabon
by codeMantra

Contents

About the editors

Amit Bage works as an assistant professor in the Department of Electronics and Communication Engineering at the National Institute of Technology Hamirpur, Himachal Pradesh, India. In his area of specialization, he is interested in electromagnetics, and his recent research activities have focused on microwave antennas, microwave passive components, and sensors. He has published more than 40 research papers in journals and conferences of national and international reputation.

Aman Kumar works as an assistant professor in the Department of Electronics and Communication Engineering at the National Institute of Technology Hamirpur, Himachal Pradesh, India. His research area includes designing filters, filter banks and their applications, biomedical signal/image processing, machine learning, deep learning algorithms for biomedical signals, and image processing. He has published more than 25 research papers in reputed journals and conferences.

Prakash Pareek is currently working as an associate professor in the Department of Electronics and Communication Engineering at Vishnu Institute of Technology, Andhra Pradesh, India. His research interests are optical communications, group IV optoelectronic devices, plasmonics, integrated optics, and THz photonics. He has published more than 70 research papers in reputed peer-reviewed journals and conferences. He is a senior member of IEEE and a professional member of Optica (Formerly Optical Society of America).

Dharmendra Prasad Mahato works as an assistant professor in the Department of Computer Science and Engineering at the National Institute of Technology Hamirpur, Himachal Pradesh, India. His research interests include distributed computing, artificial intelligence, operating systems, databases, and modeling and simulation. He has published in journals such as *Applied Soft Computing*, *Swarm and Evolutionary Computation*, *ISA Transactions*, *Cluster Computing*, *Concurrency*, and *Computation: Practice and Experience* and conferences such as AINA, ICPP, ICDCN, and E-Science.

Rardchawadee Silapunt works as an associate professor in the Department of Electronics and Telecommunication Engineering at King Mongkut's University of Technology Thonburi, Thailand. Her research interests include the Internet of Things (IoT) solutions for smart farming, asset tracking, and vehicle-to-everything. She has published more than 100 research papers in reputed journals and conferences. She is a member of IEEE and vice chair of the electromagnetics area under the Electrical Engineering/Electronics, Computer, Telecommunications, and Information Technology Association, Thailand.

Contributors

Sudhakar Ajmera
Department of Electronics and Communication Engineering
MLR Institute of Technology
Hyderabad, India

Manjunatha Reddy A. H.
Department of Biotechnology
RV College of Engineering
Bangalore, India

Venkata Naga Rani Bandaru
Department of Information Technology
Vishnu Institute of Technology, Bhimavaram
Andhra Pradesh, India

Kaushik Chaudhari
School of Electronics and Telecommunication Engineering
MIT Academy of Engineering
Alandi, Pune, India

Srikanth Cherukuvada
Department of Computer Science & Engineering (AI&ML)
St. Martin's Engineering College
Dhulapally, Secunderabad, Telangana, India

Shreya Dahatonde
School of Electronics and Telecommunication Engineering
MIT Academy of Engineering
Alandi, Pune, India

Sumit Kumar Jindal
School of Electronics Engineering
Vellore Institute of Technology
Vellore, Tamil Nadu, India

Shreya Girish Karkera
Department of Biotechnology
RV College of Engineering
Bangalore, India

Idamakanti Kasireddy
Department of Electrical Communication and Engineering
Vishnu Institute of Science and Technology
Bhimavaram, Andhra Pradesh, India

Aman Kumar
Department of Electronics and Communication Engineering
National Institute of Technology
Hamirpur, Himachal Pradesh, India

Bittu Kumar
Department of Electronics and Communication Engineering
Koneru Lakshmaiah Education Foundation
Hyderabad, India

Brajesh Kumar
Department of Electronics and Communication Engineering
Government Engineering College
Jamui, Bihar, India

Neeraj Kumar
Department of Electrical and Electronics Engineering
Galgotias University
Uttar Pradesh, India

Rahul Kumar
School of Computer Science
University of Petroleum and Energy Studies
Dehradun, Uttarakhand, India

Tatpar Kunghadkar
School of Electronics and Telecommunication Engineering
MIT Academy of Engineering
Alandi, Pune, India

Sumathra Manokaran
Department of Biotechnology
RV College of Engineering
Bangalore, India

M. D. Yaana Muthamma
Department of Biotechnology
RV College of Engineering
Bangalore, India

Savita Pawar
School of Electronics and Telecommunication Engineering
MIT Academy of Engineering
Alandi, Pune, India

Sanapathi Venkata Ashok Durga Prasad
Department of CS&SE
Andhra University college of Engineering(A)
Visakhapatnam, India

Shaik Mohammad Rafee
Department of Artificial Intelligence and Machine Learning
Sasi Institute of Technology & Engineering
Tadepalligudem, Andhra Pradesh, India

Srijeev R. Rajesh
Centre of Excellence in Computational Genomics
RV College of Engineering
Bangalore, India

B. Mohan Rao
Department of Electronics and Communication Engineering
Koneru Lakshmaiah Education Foundation
Hyderabad, Telangana, India

G. Ratnakanth
Department of Information Technology
Shri Vishnu Engineering, College for Women
Bhimavaram, Andhra Pradesh, India

Dipti Y. Sakhare
Department of Electronics and Telecommunication Engineering
MIT Academy of Engineering
Alandi, Pune, Maharashtra, India

Abhishek Sharma
National Institute of Technology Hamirpur
Hamirpur, Himachal Pradesh, India

Pooja Sharma
Department of Computer Science and Engineering
National Institute of Technology Hamirpur
Hamirpur, Himachal Pradesh, India

Shivani Sharma
Thapar Institute of Engineering and Technology
Patiala, India

R. Silapunt
Electronic and Telecommunication Engineering
King Mongkut's University of Technology Thonburi
Thonburi, Thailand

Anil Singh
Thapar Institute of Engineering and Technology
Patiala, India

Shweta Singh
Department of Computer Science and Engineering (AI & ML)
School of Engineering Dayananda Sagar University
Bengaluru, India

M. Srikanth
Department of Information Technology
SRKR Engineering College
Bhimavaram, Andhra Pradesh, India

Pilli Suneetha
Department of Information Technology
Sagi Rama Krishnam Raju Engineering College
Bhimavaram, Andhra Pradesh, India

Vikrant Verma
Department of Electronics and Telecommunication Engineering
MIT Academy of Engineering
Alandi, Pune, Maharashtra, India

Muthke Vivek
School of Electronics Engineering
Vellore Institute of Technology
Vellore, Tamil Nadu, India

Avani Vyas
Dr. B. R. Ambedkar National Institute of Technology Jalandhar
Punjab, India

Vipin Prakash Yadav
Department of Electronics and Telecommunication Engineering
MIT Academy of Engineering
Alandi, Pune, Maharashtra, India

Chapter 1

Some observations on precision agriculture using UAVs

Pooja Sharma

1.1 INTRODUCTION

Agriculture and farming emerged as a major source of food production and employment generation for thousands of years. The agriculture industry not only provides food for billions of people worldwide but also serves as the foundation of the domestic economy, with a substantial contribution to the majority of countries' GDP [20]. Nonetheless, given the world population's rapid rise, which is predicted to grow to nearly 9 billion people by 2050, significant obstacles are being faced in agricultural systems. This is due to the growing demand for land, and we also know that the agricultural land is shrinking day by day due to the rise in population. In this field of research, the growing demand for food and the shortage of land need comprehensive research so that in the future, we can produce food more efficiently. With the help of technologies such as Internet of Things (IoT), artificial intelligence (AI), and unmanned aerial vehicles (UAVs), we need to design and implement hardware and software in agriculture. We can call it precision agriculture (PA).

1.2 METHODS USED IN PA

PA makes use of numerous techniques and tools to improve agricultural operations and boost productivity. A lot of tools and techniques are used in PA such as Global Positioning System (GPS), Geographic Information Systems (GIS), Variable Rate Technology (VRT), AI, and machine learning (ML). The key methods used in PA are Aerial Imaging and Mapping, Yield Health Navigation, Loam and Pitch Investigation, Variable Rate Application (VRA) Provision, Vegetal Count and Development Examination, Unwanted Plant and Vexation Finding, water intensive care, Crop Assessment and Predicting, 3D Topography Model, and Spray and Planting (Argo-Drones). In PA for aerial image and mapping, RGB image, multispectral image, thermal image, and LiDAR (Light Detection and Ranging) are used for the detection of yield strength, wild plant growth, ground conditions,

DOI: 10.1201/9781003589730-1

Table 1.1 Hardware/methods/setup used in precision agriculture

References	*Hardware setup or methods*	*Various types of precision agriculture*
[8,6,22]	Global Positioning System (GPS)	Satellite-based, land-based signaling
[64,69,23]	Variable Methods	Data collection by maps, sensors, GPS
[71,28,57,54,4]	Yield Monitoring and Mapping	GPS to measure moisture content, seed quality, fertilizer
[26,67]	Remote Sensing	IR images, season-related variability
[67]	Unmanned Aerial Vehicles (UAVs)	Satellite, signals
[13,40]	Geographic Information System	Satellite images, aerial photos, multiple farm management
[10]	Drones	Multispectral, thermal, hyperspectral soil analysis
[32,45,15,53]	OTG-SENSOR	Monitoring of field and weather condition
[16,39,42,17]	Decision support system (DSS)	Agricultural data by real-time analysis

water pressure, irrigation leakages, or vegetal illnesses. Yield health can be navigated by Normalized Difference Vegetation Index (NDVI) and Soil-Adjusted Vegetation Index (SAVI). Earth and field investigation can quickly assess soil variability, erosion risks, and topography. VRA delivers data for prescription maps, which helps tractors and irrigators to smear water, composts, and insecticides only where it is required. Vegetal count and growth analysis involve counting plants and estimating their concentration, useful for measuring germination rates. Unwanted plant and vegetation detection used for spot-spraying using drone-based or ground-level VRA systems limits the chemical use. Thermal sensors are used in PA for classifying over-irrigated or under-irrigated areas. For crop assessment and prediction, drones are used for monitoring biomass collection and canopy cover, which helps evaluate crop health and yield potential. 3D topography model produces digital elevation models (DEMs) for water drain plan and land leveling.

In the field of PA, we see several methods proposed in literature as shown in Table 1.1.

1.3 LITERATURE SURVEY BASED ON VARIOUS UAVS USED IN PA

We have observed the numerous types of UAVs used in PA, which are summarized in Table 1.2. In the field of agriculture, several studies focus on

Table 1.2 Literature survey based on various UAVs used in precision agriculture

References	*Type of UAVs*	*Name of the crop*	*Camera type*
[8,42,2,6, 24,49,57]	Multi-rotor, autonomous UAVs	Rice	RGB multispectral, Robot Operating System (ROS) for ArUco
[22,19,71,64]	Multi-rotor, UAV-based multispectral	Maize	
[16,51,20,53, 13,21,69,58]	Multi-rotor (quadrotor), remote sensing technologies, vertical take-off and landing, drone MCDM, Harpia	All crops	GPS/GNSS, RGB, multispectral, hyperspectral
[38,39,68, 26,65,47]	Multi-rotor, remote sensing, IoT, UAV, UAS, sensor, UAV Static IPs	Vegetarian	Optical satellite data Landsat-8, high spatiotemporal resolution image, RGB visible light sensors, multispectral, hyperspectral and thermal sensors, DJI P4-multispectral, hyperspectral imaging, Pix4Dmapper, EO camera
[41,42]	3D LiDAR GPS	Crop detection with YOLOv7	ZED 2i stereo camera
[17,40,35,10]	Swarm UAVs (DAAS), GPS, GIS	Any crops	Remote sensing imagery hyperspectral, ultrasonic image sensor, Laser Range Finder sensor, RGB coloured images, laser, stereophonic camera, encoders, and GPS
[2,9,43,37]	UAV-aided MEC	Smart agriculture	Live streaming/analysis, surveillance, or environmental monitoring
[52]	Unmanned aerial systems (UAS), Quadcopter	Any crop	Multispectral imaging, image geo-referencing, RGB sensor

(*Continued*)

Table 1.2 (Continued) Literature survey based on various UAVs used in precision agriculture

References	*Type of UAVs*	*Name of the crop*	*Camera type*
[62,28,66,41]	Quadcopter from DJI, Active sensing UAV, Fly Ad hoc Network, Rotocopter airframes	Weed and crop	RGB, HSV, HSL image
[46]	Multi-rotor	Potato	Multispectral cameras, Sequoia, Sentera, ADC Micro, and Mini MCA6
[7,27,44]	Multispectral, UAV remote sensing	Geranium plant, saffron fields, AFarCloud, DEMETER, Landsat-8	RGB Olympus PEN E-PM1
[36,34,72]	UAV remote sensing, multispectral	Moreira 2021 precision, Landsat-8, Sentinel-2, Planet Scope	RGB Olympus PEN E-PM1
[25]	Autonomous multisensor UAV	Pomegranate	Multispectral image, thermal image
[18]	OBAM	Species plants	RGB
[32,70]	Multi-efficient channel, multi-temporal	Wheat, wheat ears	Yolov8-N
[45]	Multi-temporal	Aerial crop	NDVI cameras on drones
[12,15,55,11]	IoT devices with fix sensors, actuators, palm UAVs	Vegetation and crop	Solar-reflected optical spectrum (400–2500 nm), Maple leaf
[67]	Flying Ad hoc Network	Precision agriculture	UAV with on-board cameras
[60]	AIoT UAV-based smart sensor	Agriculture	AIoT sensors
[50]	Multisensor hierarchical UAV swarm	Soybean	RGB, multispectral, thermal
[33]	IoT, remote sensing, unmanned aerial vehicles	Smart agriculture	Camera, NIR

(*Continued*)

Table 1.2 (Continued) Literature survey based on various UAVs used in precision agriculture

References	*Type of UAVs*	*Name of the crop*	*Camera type*
[1]	High spatial resolution	Grid-based soil sampling for crops	Sensors
[73]	Tank-mix adjuvants using UAVs	Rice	Optical image
[74]	U-Net	Spatial segregation and segmentation	RGB + NIR
[59]	UAV-PA, UAV-PV	Agricultural and forest meteorology	Remotely piloted aircraft
[29]	On-board UAVs	Vegetation	DJI Phantom-4
[31]	Optical remote Sensing, ground control points	Olive plant	RGB, multispectral, hyperspectral, thermal
[61]	Multi-rotor UAV	Multiple crops	WSN-UAV
[5]	Quadcopter UAVs	Trees	RGB, multispectral, hyperspectral, LiDAR
[30]	Remote sensing technology	Broccoli	Visible/multispectral UAV image
[54,4,63]	Remote sensing	Sugarcane	RGB, multispectral, hyperspectral, LiDAR, thermal cameras
[14]	RTK-GNSS GPS	Rose	—
[23]	DJI Phantom 4 Pro-Drone	Spinach	Multispectral UAV

various types of grains which are important for today's population, such as rice, wheat, maize, sugarcane, potato, vegetables, pulses, fruits, and many more. In rice cultivation, several techniques have been used such as multi-rotor, autonomous UAVs, RGB multispectral imaging, and Robot Operating System (ROS) for ArUco [8,42,2,6,24,49,57]. For the cultivation of maize, multi-rotor, UAV-based multispectral imaging has been used [22,19,71,64]. Multi-rotor (quadrotor) UAVs, remote sensing technologies, vertical take-off and landing systems, drone-based multi-criteria decision-making (MCDM), and Harpia platforms have been used for all types of crops using GPS/GNSS, RGB, multispectral data, and hyperspectral data [16,51,20,53,13,21,69,58]. Multi-rotor, remote sensing, IoT, UAV, unmanned aerial systems (UAS), sensor, UAV static IPs have been used for vegetation and optical satellite data Landsat-8, High spatio-temporal resolution image, Visible light sensors RGB, Multispectral sensors, Hyperspectral sensor, Thermal sensors, Da-Jiang Innovation (DJI) P4-multispectral, Hyperspectral imaging, Pix4Dmapper, EO camera have been used for high-resolution agricultural monitoring [39,38,68,26,65,47]. 3D LiDAR GPS is used for crop detection using YOLOv7 and ZED 2i Stereo Camera [42]. In Refs. [17,40,35,10], swarm UAVs (DAAS), GPS, GIS have been used for all types of crops using remote sensing imagery, hyperspectral RGB images, LRF sensor, ultrasonic, laser scanner, stereo camera, encoders, and GPS. In Refs. [2,9,43,37], UAV-aided mobile edge computing (MEC) has been used for smart agriculture by using live streaming/analysis, surveillance, or environmental monitoring. In Ref. [52], UAS and Quadcopter have been used for all types of crops using multispectral imaging, image geo-referencing, and RGB sensor. In Refs. [62,28,66,41], quadcopter from DJI, active sensing UAVs, Flying Ad hoc Network, and Rotocopter airframes have been used for weed and crop monitoring using RGB, HSV, and HSL image data. In Ref. [46], the multi-rotor technique has been used for potato monitoring using multispectral cameras, Sequoia, Sentera, ADC Micro, and Mini-MCA6. In Refs. [7,27,44,36,34,72], multispectral and UAV remote sensing have been used for checking Geranium plant, saffron fields using Afar Cloud, DEMETER, moreira2021precision, Landsat-8, Sentinel-2, Planet Scope, and RGB Olympus PEN E-PM1. Autonomous multisensor UAV has been used for pomegranate plants by using multispectral image and thermal image [25]. In Ref. [18], OBAM technique has been used for many plant species using RGB. Multi-channel and multi-temporal approaches have been implemented for wheat and wheat ear detection using Yolov8-N [32,70], and multi-temporal analysis has been used for aerial crop using NDVI cameras mounted on drones [45]. IoT devices with fixed sensors, actuators, and palm UAVs have been used for vegetation and crop monitoring using solar-reflected optical spectrum (400–2500 nm) and maple leaf [12,15,55,11]. Flying Ad hoc Network has been used in PA using UAV with on-board cameras [67], and AIoT UAV-based smart sensor has been used in agriculture using

AIoT sensors [60]. In Ref. [50], multisensor hierarchical UAV swarm has been implemented for soybean monitoring using RGB, multispectral and thermal imaging. In Ref. [33], IoT, remote sensing, and UAVs have been used for smart agriculture with camera and near infrared. In Ref. [29], on-board UAVs have been used for vegetation using DJI Phantom-4. Multi-rotor UAVs have been used for multiple crops using WSN-UAV [61]. Quadcopter UAVs have been used for monitoring of trees, integrating RGB, multispectral, hyperspectral, and LiDAR sensors [5]. High spatial resolution setup has been used for grid-based soil sampling for crops using sensors [1]. In Ref. [73], tank-mix adjuvants using UAVs have been implemented for rice using optical image. UAV-PA and UAV-PV for agricultural and forest meteorology have been used with remotely piloted aircraft [59]. Optical remote sensing and ground control points have been used for olive plants using RGB, multispectral, hyperspectral, and thermal imaging [31]. In Ref. [30], remote sensing technology has been used for broccoli with visible/multispectral UAV image. Remote sensing has been used for sugarcane with RGB, multispectral, hyperspectral, LiDAR data using thermal cameras [54,4,63]. In Ref. [14], RTK-GNSS GPS has been used for rose plants. In smart agriculture, IoT, remote sensing, and UAV technologies are used with RGB, NIR images, and image datasets [33]. Sugarcane crop [54,4,63] is maintained by remote sensing and their images are maintained by RGB, multispectral and hyperspectral imaging, LiDAR, and thermal cameras. Pro-Drone and DJI Phantom-4 technology is used to maintain spinach crops using multispectral UAVs [23].

1.4 DATASETS USED IN PA

In PA, various datasets are used to collect information regarding soil conditions, crop types, weather prediction, and other environmental factors. These datasets help us to analyze the real-time, high-resolution, and actionable data. These datasets authorize growers, agronomists, and investigators to optimize field management, improve crop yields, reduce environmental impact, and lower input costs. For multispectral image, specific frequency bands across the electromagnetic spectrum with multispectral sensors are used. Agri-FoRe and PhenoFly datasets are examples of multispectral images. Hyperspectral dataset having hundreds of narrow spectral bands detects subtle changes in crop composition and contain HyPlant UAV dataset, WHU-Hi-A dataset. RGB image datasets contain OpenAerialMap Agricultural RGB dataset and UAVid dataset. Thermal imaging datasets, such as FLIR thermal dataset for UAV-based agricultural irrigation efficiency thermal datasets, are used for monitoring irrigation performance. These datasets are collected using resources like sensors, satellites, drones, IoT, AI, ML, and ground-based measurements. This section discusses the datasets used in the field of PA (seen in Table 1.3).

Table 1.3 Dataset used in precision agriculture

References	*Data type*	*Class of data*
[39]	Landsat-8 data with 11 spectral bands	Sparse and dense vegetation
[42]	Yolov7: Wild and cultivated plants 2150 images	Different vegetation and crops and weeds
[23]	Vegetation indices (VIs), Da-Jiang Innovation (DJI) Phantom 4 Drone Quadcopter UAV with 20MP, 1" CMOS sensor 4K H.265 Video 3-Axis, Remote Controller having 5.5" White (CP.PT.00000234.01)	Spinach field
[64]	SI-111 IRR maize temperature data	Maize field
[14]	Sentara sensor on the UAV platform	Organic oil rose farming
[54]	Lapix dataset with RGB images of 256*256 pels	Sugarcane crop
[57,38,58]	MADA and Beras Nasional (BERNAS)	Rice
[28]	Weed map	Sugar beet crops and weeds interfering
[71]	YOLOv5, OTSU	Corn plants
[35]	RGB pest images dataset with 40 images	Vegetation and pest detection
[27]	Orthophotos in the RGB and NDVI spectrum	Saffron, weed, field, mammal sample
[26]	Leaf spectra datasets	Vegetation indices
[48]	Pix4Dmapper for time-series database	Crop health surveillance, vineyard pests and diseases
[34]	Landsat-8, Sentinel-2, Planet Scope	All crops with the occurrence of weeds and identification of plant parts affected by water/nitrogen pressure
[32]	Yolov8-UAV, CBDA, and WEDU datasets	Maize braids identification and count (MTDC), RFRB, CBDA, WEDU
[56]	Deep potato	Potato (Lady Claire and Markies)
[3]	CHM–NDVI data	Orchard tree

1.5 CHALLENGES AND OPEN PROBLEMS IN PA USING UAVs

Today's PA is a dynamic and flexible technology. PA is influenced by various technologies, like UAVs, biotechnology, drones, blockchain, big data, and AI. We can see the details of challenges and open problems in Table 1.4.

Table 1.4 Observation observed in precision agriculture using UAV technologies

References	*Observations*
[8,6,22]	To abate the expenditure of rice during its production and processing.
[16,20,39]	Some machine learning algorithms (Double Deep Q-Learning, Reinforcement learning) can also be used for novel Global Mission Planner (GMP).
[42]	The study can be tested in real environment scenario.
[17]	Novel security measures like encryption methods can be improved by the use of FL.
	For the development of a large system, the token system can be generalized for tasks such as renting different drone models, refining data collection methods with numerous drones, collecting samples, and transporting yields to storage or sale.
[10]	To explore the viability of uniting the drone with a classical scheme, sense-store-forward. To study aeronautical approaches for the drone in the situation of multiple mixed data collection manner.
[2,7,21]	Dynamic location and development of Network Function Virtualization-enabled in a farm mechanization application by minimizing the energy consumption in the network checked for real enlivenment scenario.
[52]	To develop orthomosaic dataset for the vegetation or soil structure, the pre-eminent snapshots are moved to a set of snap expressions and joined with explicit metadata. The vegetation record will be received from the closure view of infrared orthomosaics to deliver farmers an accurate and precise picture of the vegetation type.
[37]	Prevention of drone crashes by "separating" routes (deviation of their routes).
[24]	The project can be enhanced by implementing a UAV in a farmhouse and a smart IoT-based hydroponic farmhouse for virus detection using digital image processing.
[46]	The study can be extended for different crop fields and spectral indices like effect of weather fluctuations and flight elevation on the adjustment of the camera.
[18,25,26,27,65]	For future plan, we need standardized reduced-input applications and services by accentuating the possibilities of full robotics in multidirectional spectral data gathering with UAVs.
[66]	To propose a framework for evaluating the drone network performance.
[15]	For future research, optimized intervention techniques need to be proposed to reduce pesticide use and costs.
[5,49,61,19,62]	For future work, more efficient algorithms and spraying methods are to be developed, which helps the growers to raise production.
[53,60,55,45,32]	For future work, the drones can be enabled with innovative technologies like image dispensation, visual systems, data analytics, artificial intelligence algorithms, efficient spraying method which helps the growers to increase production and make agriculture more precise and efficient.

(*Continued*)

Table 1.4 (Continued) Observation observed in precision agriculture using UAV technologies

References	*Observations*
[34]	For future works, more tests will be performed on multiple seasons and also extend the test area.
[11,12,67]	For future work, the model can be compared with path loss and also validated for its accuracy.
[50]	For future research to improve swarm performance, scale, and adoption for smart agriculture, new technologies can be explored.
[72]	The work can be extended by agricultural communities by implementing data analytics and farming as a service (FaaS) by performing continuous development of production processes.
[33,36,43,47, 1,13,73,31]	For future work, some specific necessities and tasks need to be handled before UAVs can be made functional for smart farming implementation.
[41,59,51]	For future research, there needs to be various technological issues like improvement in battery life in UAV platforms and fusion of colour, spectral and thermal data products for the improvement of reliability and accuracy.
[40]	For future work, sustainable precision agriculture can be done for technological innovation, sustainable farming practices, data analytics, and policy interventions.
[35]	For future work, algorithm needs to be developed to reduce the overdose of toxic pesticides.
[74]	Energy-aware and low-cost models need to be developed in the future which will be organized on the UAVs to achieve real-time, quicker descriptions in the area of precision agriculture.
[68]	For future research, weed mapping like OBIA using machine learning techniques should be done.
[29]	For future work, the computational complexity may be compensated by using a high-speed computing platform.
[44]	Using spatiotemporal semantic data management tool, the precipitation amounts or wind speed and direction can be measured.
[38]	Some new approaches integrated with the UAVs can be explored for gaining efficiency in precision agriculture.
[58]	For future research, the application of rice paddy yield for biomass construction and power generation needs to be enhanced which will reduce smog index in the air and offer innovative waste management solutions.
[30,57]	Future work needs to optimize the irrigation and fertilization decision support system to improve the effectiveness of multispectral UAV images joined with DL and object detection and semantic segmentation for retrieving other crop images.
[71,28]	For further work, YOLOv5 methods can be generalized to progress agriculture procedures for other yields.
[54,63,14,9]	Future work includes developing an API (Application Programming Interface) for storing the location in the dataset and performing intermittent monitoring for the changes in the crowded zones using temporal analysis.

(*Continued*)

Table 1.4 (Continued) Observation observed in precision agriculture using UAV technologies

References	*Observations*
[64]	For future research, more types of yield can be examined using PrA technology.
[69,23]	Further plan validation of data collected from various sensor technologies needs to be joined with BN and GPS for refining safety and accountability.

- They may also create new opportunities for collaboration, transparency, and innovation along the whole agri-food value chain. These technologies do, however, also bring with them new hazards and uncertainties, which call for regulation.
- Coordination among the farmers because of quite expensive PA equipment such as GIS, GPS, RS, etc. is needed. Farmers need to invest in apparatus and software technology used for PA and learn to use PA equipment, which may not be affordable or accessible for many in developing countries like India.
- One of the key challenges is the high price and multifaceted technology involved.
- To explore the viability of integrating the use of the drone with the sense-store-forward communication system. To study drone flight approaches for different data collection scenarios.
- Future work needs to focus on integrating IoT technology with UAV systems and ML algorithms to usher in a new era for the agri-food industry.
- To develop orthomosaics with accurate information on the vegetation and soil conditions, aerial images are processed to remove distortions and incorporate precise metadata. The plant data are then extracted from visible and near-infrared (NIR) orthomosaics to provide growers with a more exact representation of crop conditions.
- Work can be extended using UAV cloud platform in applications such as industrial automation, aerial surveillance, civil construction, and building maintenance with multiple UAVs.

1.6 CONCLUSION

In this chapter, we tried to have some observations on the topic of PA. We have also discussed about the datasets used in the literature. We have observed several types of UAVs and IoT devices used in this research area. The merging of UAVs with PA makes a noteworthy development in how modern farming techniques succeed and have improved crops, soil, and weather conditions. A lot of challenges are involved in worldwide food security, weather variation, lack of labour, and resource exhaustion which need advanced and effective techniques, and UAV-based technologies have emerged as one of the most promising tools to address these complex issues. PA emphasizes site-specific management, real-time monitoring, and data-driven decisions, aiming to increase crop productivity, reduce resource wastage, and improve sustainability. UAVs play a central role in achieving these objectives by providing accurate, timely, and scalable insights into farm conditions. Equipped with advanced sensors and imaging systems, drones allow farmers to gain aerial perspectives of their fields with unprecedented clarity and detail. This shift from manual, generalized farming methods to precise, technology-enabled solutions is fundamentally transforming the agricultural sector. The benefits of UAVs in PA span multiple dimensions. The chapter also discusses the latest algorithms of deep learning and ML which have been implemented so far. Finally, we have observed some important open problems that have not been solved so far.

REFERENCES

1. Ackerson, J.P.: Precision agriculture and soil health. *Crops & Soils* 55(5), 35–39 (2022).
2. Akbari, M., Syed, A., Kennedy, W.S., Erol-Kantarci, M.: Constrained federated learning for AoI-limited SFC in UAV-aided MEC for smart agriculture. *IEEE Transactions on Machine Learning in Communications and Networking* 1, 277–295 (2023).
3. Altieri, G., Maffia, A., Pastore, V., Amato, M., Celano, G.: Use of high-resolution multispectral UAVs to calculate projected ground area in *Corylus avellana* l. Tree orchard. *Sensors* 22(19), 7103 (2022).
4. Amarasingam, N., Salgadoe, A.S.A., Powell, K., Gonzalez, L.F., Natarajan, S.: A review of UAV platforms, sensors, and applications for monitoring of sugarcane crops. *Remote Sensing Applications: Society and Environment* 26, 100712 (2022).
5. Ampatzidis, Y., Partel, V., Costa, L.: Agroview: Cloud-based application to process, analyze and visualize UAV-collected data for precision agriculture applications utilizing artificial intelligence. *Computers and Electronics in Agriculture* 174, 105457 (2020).
6. An, G., Xing, M., He, B., Kang, H., Shang, J., Liao, C., Huang, X., Zhang, H.: Extraction of areas of rice false smut infection using UAV hyperspectral data. *Remote Sensing* 13(16), 3185 (2021).
7. Bagha, H., Yavari, A., Georgakopoulos, D.: IoT-based plant health analysis using optical sensors in precision agriculture. In: *2021 Digital Image Computing: Techniques and Applications (DICTA)*. pp. 01–08. IEEE, Gold Coast, Australia (2021).
8. Bhoi, S.K., Jena, K.K., Panda, S.K., Long, H.V., Kumar, R., Subbulakshmi, P., Jebreen, H.B.: An internet of things assisted unmanned aerial vehicle based artificial intelligence model for rice pest detection. *Microprocessors and Microsystems* 80, 103607 (2021).

9. Boursianis, A.D., Papadopoulou, M.S., Diamantoulakis, P., Liopa-Tsakalidi, A., Barouchas, P., Salahas, G., Karagiannidis, G., Wan, S., Goudos, S.K.: Internet of things (IoT) and agricultural unmanned aerial vehicles (UAVs) in smart farming: A comprehensive review. *Internet of Things* 18, 100187 (2022).
10. Caruso, A., Chessa, S., Escolar, S., Barba, J., López, J.C.: Collection of data with drones in precision agriculture: Analytical model and lora case study. *IEEE Internet of Things Journal* 8(22), 16692–16704 (2021).
11. Castellanos, G., Deruyck, M., Martens, L., Joseph, W.: System assessment of WUSN using NB-IOT UAV-aided networks in potato crops. *IEEE Access* 8, 56823–56836 (2020).
12. Cavaliere, D., Loia, V., Senatore, S.: Towards a layered agent-modeling of IoT devices to precision agriculture. In: *2020 IEEE International Conference on Fuzzy Systems (FUZZ-IEEE)*. pp. 1–8. IEEE, Glasgow, UK (2020).
13. Degieter, M., De Steur, H., Tran, D., Gellynck, X., Schouteten, J.J.: Farmers' acceptance of robotics and unmanned aerial vehicles: A systematic review. *Agronomy Journal* 115(5), 2159–2173 (2023).
14. Demir, S., Dedeoğlu, M., Başayiğit, L.: Yield prediction models of organic oil rose farming with agricultural unmanned aerial vehicles (UAVs) images and machine learning algorithms. *Remote Sensing Applications: Society and Environment* 33, 101131 (2024).
15. Geckeler, C., Ramos, S.E., Schuman, M.C., Mintchev, S.: Robotic volatile sampling for early detection of plant stress: Precision agriculture beyond visual remote sensing. *IEEE Robotics & Automation Magazine* 30, 41–51 (2023).
16. Gopalakrishnan, S.K., Al-Rubaye, S., Inalhan, G.: Adaptive UAV swarm mission planning by temporal difference learning. In: *2021 IEEE/AIAA 40th Digital Avionics Systems Conference (DASC)*. pp. 1–10. IEEE, San Antonio, TX (2021).
17. Guven, I., Parlak, M.: Blockchain, AI and IoT empowered swarm drones for precision agriculture applications. In: *2022 IEEE 1st Global Emerging Technology Blockchain Forum: Blockchain & Beyond (iGETblockchain)*. pp. 1–6. IEEE, Irvine, CA (2022).
18. Huang, H., Tang, Y., Tan, Z., Zhuang, J., Hou, C., Chen, W., Ren, J.: Object-based attention mechanism for color calibration of UAV remote sensing images in precision agriculture. *IEEE Transactions on Geoscience and Remote Sensing* 60, 1–13 (2022).
19. Ishengoma, F.S., Rai, I.A., Said, R.N.: Identification of maize leaves infected by fall armyworms using UAV-based imagery and convolutional neural networks. *Computers and Electronics in Agriculture* 184, 106124 (2021).
20. Istiak, M.A., Syeed, M.M., Hossain, M.S., Uddin, M.F., Hasan, M., Khan, R.H., Azad, N.S.: Adoption of unmanned aerial vehicle (UAV) imagery in agricultural management: A systematic literature review. *Ecological Informatics* 78, 102305 (2023).
21. Jasim, A.N., Fourati, L.C., Albahri, O.: Evaluation of unmanned aerial vehicles for precision agriculture based on integrated fuzzy decision-making approach. *IEEE Access* 11, 75037–75062 (2023).
22. Jia, X., Yin, D., Bai, Y., Yu, X., Song, Y., Cheng, M., Liu, S., Bai, Y., Meng, L., Liu, Y., et al.: Monitoring maize leaf spot disease using multi-source UAV imagery. *Drones* 7(11), 650 (2023).

23. Jiang, X., Sun, S., Adamchuk, V., Weisman, G., Ramsay, P., Rincon, A.R.: Spinach yield mapping using multispectral UAV imagery. *Smart Agricultural Technology* 7, 100389 (2024).
24. Jibon, Z.A., Adnan, M.A., Nora, N.T., Akash, M.M.H., Ahammed, F.: Development of an autonomous UAV for seed and fertilizer distribution in precision agriculture. In: *2023 14th International Conference on Computing Communication and Networking Technologies (ICCCNT)*. pp. 1–5. IEEE, Delhi, India (2023).
25. Katsigiannis, P., Misopolinos, L., Liakopoulos, V., Alexandridis, T.K., Zalidis, G.: An autonomous multi-sensor UAV system for reduced-input precision agriculture applications. In: *2016 24th Mediterranean Conference on Control and Automation (MED)*. pp. 60–64. IEEE, Athens, Greece (2016).
26. Khalesi, F., Daponte, P., De Vito, L., Picariello, F., Tudosa, I.: UAV in precision agriculture: A preliminary assessment of uncertainty for vegetation health index. In: *2022 IEEE Workshop on Metrology for Agriculture and Forestry (MetroAgriFor)*. pp. 94–99. IEEE, Perugia, Italy (2022).
27. Kiropoulos, K., Dimaraki, F., Tsouros, D.C., Bibi, S.: UAV saffran monitoring process. In: *2022 11th International Conference on Modern Circuits and Systems Technologies (MOCAST)*. pp. 1–4. IEEE, Bremen, Germany (2022).
28. Krestenitis, M., Raptis, E.K., Kapoutsis, A.C., Ioannidis, K., Kosmatopoulos, E.B., Vrochidis, S.: Overcome the fear of missing out: Active sensing UAV scanning for precision agriculture. *Robotics and Autonomous Systems* 172, 104581 (2024).
29. Latif, M.A.: Improving stability of aerial videos acquired through vision sensors onboard UAVs for applications in precision agriculture. *Signal, Image and Video Processing* 16(5), 1263–1270 (2022).
30. Lee, C.J., Yang, M.D., Tseng, H.H., Hsu, Y.C., Sung, Y., Chen, W.L.: Single-plant broccoli growth monitoring using deep learning with UAV imagery. *Computers and Electronics in Agriculture* 207, 107739 (2023).
31. López, A., Jurado, J.M., Ogayar, C.J., Feito, F.R.: A framework for registering UAV-based imagery for crop-tracking in precision agriculture. *International Journal of Applied Earth Observation and Geoinformation* 97, 102274 (2021).
32. Lu, D., Ye, J., Wang, Y., Yu, Z.: Plant detection and counting: Enhancing precision agriculture in UAV and general scenes. *IEEE Access* 11, 116196–116205 (2023).
33. Maddikunta, P.K.R., Hakak, S., Alazab, M., Bhattacharya, S., Gadekallu, T.R., Khan, W.Z., Pham, Q.V.: Unmanned aerial vehicles in smart agriculture: Applications, requirements, and challenges. *IEEE Sensors Journal* 21(16), 17608–17619 (2021).
34. Mancini, A., Frontoni, E., Zingaretti, P.: Satellite and UAV data for precision agriculture applications. In: *2019 International Conference on Unmanned Aircraft Systems (ICUAS)*. pp. 491–497. IEEE, Atlanta, GA (2019).
35. Meshram, A.T., Vanalkar, A.V., Kalambe, K.B., Badar, A.M.: Pesticide spraying robot for precision agriculture: A categorical literature review and future trends. *Journal of Field Robotics* 39(2), 153–171 (2022).
36. Moreira, M., Azevedo, F., Ferreira, A., Pedro, D., Matos-Carvalho, J., Ramos, Á., Loureiro, R., Campos, L.: Precision landing for low-maintenance remote operations with UAVs. *Drones* 5(4), 103 (2021).

37. Mukhamediev, R.I., Yakunin, K., Aubakirov, M., Assanov, I., Kuchin, Y., Symagulov, A., Levashenko, V., Zaitseva, E., Sokolov, D., Amirgaliyev, Y.: Coverage path planning optimization of heterogeneous UAVs group for precision agriculture. *IEEE Access* 11, 5789–5803 (2023).
38. Mukherjee, A., Misra, S., Raghuwanshi, N.S.: A survey of unmanned aerial sensing solutions in precision agriculture. *Journal of Network and Computer Applications* 148, 102461 (2019).
39. Murugan, D., Garg, A., Ahmed, T., Singh, D.: Fusion of drone and satellite data for precision agriculture monitoring. In: *2016 11th International Conference on Industrial and Information Systems (ICIIS)*. pp. 910–914. IEEE, Roorkee, India (2016).
40. Nath, S.: A vision of precision agriculture: Balance between agricultural sustainability and environmental stewardship. *Agronomy Journal* 116, 1–18 (2023).
41. Olson, D., Anderson, J.: Review on unmanned aerial vehicles, remote sensors, imagery processing, and their applications in agriculture. *Agronomy Journal* 113(2), 971–992 (2021).
42. OrtataŞ, F.N., UlutaŞ, H., ŞahÌn, M.E., ÇÌftcÌ, F.: Autonomous mapping and spraying in precision agriculture using unmanned ground vehicles. In: *2023 Innovations in Intelligent Systems and Applications Conference (ASYU)*. pp. 1–5. IEEE, Sivas, Turkiye (2023).
43. Pamuklu, T., Syed, A., Kennedy, W.S., Erol-Kantarci, M.: Heterogeneous GNN-RL-based task offloading for UAV-aided smart agriculture. *IEEE Networking Letters* 5, 213–217 (2023).
44. San Emeterio de la Parte, M., Martínez-Ortega, J.F., Hernández Díaz, V., Martínez, N.L.: Big data and precision agriculture: A novel spatio-temporal semantic IoT data management framework for improved interoperability. *Journal of Big Data* 10(1), 1–32 (2023).
45. Patidar, P.K., Tomar, D.S., Pateriya, R., Sharma, Y.K.: Precision agriculture: Crop image segmentation and loss evaluation through drone surveillance. In: *2023 Third International Conference on Secure Cyber Computing and Communication (ICSCCC)*. pp. 495–500. IEEE, Jalandhar, India (2023).
46. Pérez-García, Á., Rodríguez-Molina, A., Hernández, E., Vera, L., López, J.F.: Development of low-cost multi-spectral cameras for precision agriculture. In: *IGARSS* 2023–2023 *IEEE International Geoscience and Remote Sensing Symposium*. pp. 3466–3469. IEEE, Pasadena, CA (2023).
47. Phang, S.K., Chiang, T.H.A., Happonen, A., Chang, M.M.L.: From satellite to UAV-based remote sensing: A review on precision agriculture. *IEEE Access* 11, 127057–127076 (2023).
48. Pino, M., Matos-Carvalho, J.P., Pedro, D., Campos, L.M., Seco, J.C.: UAV cloud platform for precision farming. In: *2020 12th International Symposium on Communication Systems, Networks and Digital Signal Processing (CSNDSP)*. pp. 1–6. IEEE, Porto, Portugal (2020).
49. Puppala, H., Peddinti, P.R., Tamvada, J.P., Ahuja, J., Kim, B.: Barriers to the adoption of new technologies in rural areas: The case of unmanned aerial vehicles for precision agriculture in India. *Technology in Society* 74, 102335 (2023).
50. Qu, C., Boubin, J., Gafurov, D., Zhou, J., Aloysius, N., Nguyen, H., Calyam, P.: UAV swarms in smart agriculture: Experiences and opportunities. In: *2022 IEEE* 18*th International Conference on e-Science (e-Science)*. pp. 148–158. IEEE, Salt Lake City, UT (2022).

51. Radoglou-Grammatikis, P., Sarigiannidis, P., Lagkas, T., Moscholios, I.: A compilation of UAV applications for precision agriculture. *Computer Networks* 172, 107148 (2020).
52. Rajapu, A., Madisetty, S.A., Thokala, S., Mahendra, P.R., Raju, V.V.R., Reddy, C.S.V.: Cost-effective multispectral imaging system for precision agriculture. In: *2022 IEEE 2nd International Conference on Sustainable Energy and Future Electric Transportation (SeFeT)*. pp. 1–8. IEEE, Hyderabad, India (2022).
53. Rakesh, D., Kumar, N.A., Sivaguru, M., Keerthivaasan, K., Janaki, B.R., Raffik, R.: Role of UAVs in innovating agriculture with future applications: A review. In: *2021 International Conference on Advancements in Electrical, Electronics, Communication, Computing and Automation (ICAECA)*. pp. 1–6. IEEE, Coimbatore, India (2021).
54. Ribeiro, J.B., da Silva, R.R., Dias, J.D., Escarpinati, M.C., Backes, A.R.: Automated detection of sugarcane crop lines from UAV images using deep learning. *Information Processing in Agriculture* 11, 385–396 (2023).
55. Rosewick, S., Schorlemer, F., Kunemund, F.: System architecture for a wireless sensor network for sustainable agriculture, forestry and green spaces in smart cities. In: *2023 IEEE 12th International Conference on Intelligent Data Acquisition and Advanced Computing Systems: Technology and Applications (IDAACS)*. vol. 1, pp. 181–186. IEEE, Dortmund, Germany (2023).
56. Ruszczak, B., Boguszewska-Mańkowska, D.: Deep potato – The hyperspectral imagery of potato cultivation with reference agronomic measurements dataset: Towards potato physiological features modeling. *Data in Brief* 42, 108087 (2022).
57. Sah, S.S., Maulud, K.N.A., Sharil, S., Karim, O.A., Pradhan, B.: Monitoring of three stages of paddy growth using multispectral vegetation index derived from UAV images. *The Egyptian Journal of Remote Sensing and Space Sciences* 26(4), 989–998 (2023).
58. Sharma, A., Singh, P.K.: Applicability of UAVs in detecting and monitoring burning residue of paddy crops with IoT integration: A step towards greener environment. *Computers & Industrial Engineering* 184, 109524 (2023).
59. Singh, A.P., Yerudkar, A., Mariani, V., Iannelli, L., Glielmo, L.: A bibliometric review of the use of unmanned aerial vehicles in precision agriculture and precision viticulture for sensing applications. *Remote Sensing* 14(7), 1604 (2022).
60. Singh, H., Singh, M.B., Pratik, H., Pratap, A.: UAV and UGV assisted path planning for sensor data collection in precision agriculture. In: *2023 11th International Symposium on Electronic Systems Devices and Computing (ESDC)*. vol. 1, pp. 1–6. IEEE, Sri City, India (2023).
61. Singh, P.K., Sharma, A.: An intelligent WSN-UAV-based IoT framework for precision agriculture application. *Computers and Electrical Engineering* 100, 107912 (2022).
62. Singh, V., Singh, D.: Development of an approach for early weed detection with UAV imagery. In: *IGARSS* 2022–2022 *IEEE International Geoscience and Remote Sensing Symposium*. pp. 4879–4882. IEEE, Kuala Lumpur, Malaysia (2022).

63. Srivastava, K., Bhutoria, A.J., Sharma, J.K., Sinha, A., Pandey, P.C.: UAVs technology for the development of GUI based application for precision agriculture and environmental research. *Remote Sensing Applications: Society and Environment* 16, 100258 (2019).
64. Széles, A., Huzsvai, L., Mohammed, S., Nyéki, A., Zagyi, P., Horváth, É., Simon, K., Arshad, S., Tamás, A.: Precision agricultural technology for advanced monitoring of maize yield under different fertilization and irrigation regimes: A case study in eastern Hungary (Debrecen). *Journal of Agriculture and Food Research* 15, 100967 (2024).
65. Triantafyllou, A., Sarigiannidis, P., Bibi, S., Vakouftsi, F., Vassilis, P.: Modelling deployment costs of precision agriculture monitoring systems. In: *2020 16th International Conference on Distributed Computing in Sensor Systems (DCOSS)*. pp. 252–259. IEEE, Marina del Rey, CA (2020).
66. Tropea, M., Santamaria, A.F., De Rango, F., Potrino, G.: Reactive flooding versus link state routing for FANET in precision agriculture. In: *2019 16th IEEE Annual Consumer Communications & Networking Conference (CCNC)*. pp. 1–6. IEEE, Las Vegas, NV (2019).
67. Tropea, M., Serianni, A.: Bio-inspired drones recruiting strategy for precision agriculture domain. In: *2020 IEEE/ACM 24th International Symposium on Distributed Simulation and Real Time Applications (DS-RT)*. pp. 1–4. IEEE, Prague, Czech Republic (2020).
68. Tsouros, D.C., Bibi, S., Sarigiannidis, P.G.: A review on UAV-based applications for precision agriculture. *Information* 10(11), 349 (2019).
69. Vannini, V., de Moura Souza, G., Toledo, C.F.M.: Harpia: A hybrid system for agricultural UAV missions. *Smart Agricultural Technology* 4, 100191 (2023).
70. Wei, L., Yang, H., Niu, Y., Zhang, Y., Xu, L., Chai, X.: Wheat biomass, yield, and straw-grain ratio estimation from multi-temporal UAV-based RGB and multispectral images. *Biosystems Engineering* 234, 187–205 (2023).
71. Xiao, J., Suab, S.A., Chen, X., Singh, C.K., Singh, D., Aggarwal, A.K., Korom, A., Widyatmanti, W., Mollah, T.H., Minh, H.V.T., et al.: Enhancing assessment of corn growth performance using unmanned aerial vehicles (UAVs) and deep learning. *Measurement* 214, 112764 (2023).
72. Yaqot, M., Menezes, B.C.: Unmanned aerial vehicle (UAV) in precision agriculture: business information technology towards farming as a service. In: *2021 1st International Conference on Emerging Smart Technologies and Applications (eSmarTA)*. pp. 1–7. IEEE, Sana'a, Yemen (2021).
73. Zhao, R., Sun, Z., Bird, N., Gu, Y.C., Xu, Y., Zhang, Z.H., Wu, X.M.: Effects of tank-mix adjuvants on physicochemical properties and dosage delivery at low dilution ratios for unmanned aerial vehicle application in paddy fields. *Pest Management Science* 78(4), 1582–1593 (2022).
74. Zualkernan, I., Abuhani, D.A., Hussain, M.H., Khan, J., ElMohandes, M.: Machine learning for precision agriculture using imagery from unmanned aerial vehicles (UAVs): A survey. *Drones* 7(6), 382 (2023).

Chapter 2

Predictive analysis techniques for crop and soil sensors

Avani Vyas, Anil Singh, Shivani Sharma, and Abhishek Sharma

2.1 INTRODUCTION

Predictive analysis in agriculture leverages advanced technology, including sensors, machine learning (ML) algorithms, and statistical models, to analyze complex datasets. This innovative approach forecasts future crop trends, optimizes decision-making, and improves farming practices [1]. Traditional agricultural methods often relied on manual observations, intuition, and historical practices, which, while effective in some cases, lacked the precision and scalability needed to address modern challenges. The increasing demand for food production, fluctuating climatic conditions, and limited natural resources emphasize the increasing need for proactive farming tools rather than reactive decision-making.

Predictive analysis employs predictive models along with historical as well as real-time data. This allows the farmer to harvest better and in accordance with market supply. Predictive approach to smart agriculture can be enabled with the help of integration of Internet of Things (IoT) and edge computing in agricultural activities. Figure 2.1 demonstrates how sensor technology produces real-time data from farms and this is transmitted wirelessly to edge devices that are responsible for faster response. The data is further sent to cloud servers for storage and analysis. Edge and cloud devices support the predictive analysis models. Figure 2.1 outlines the widely known prediction models, based on ML, time-series analysis, image analysis, and geostatistical methods, that ensure increased productivity and sustainability in agriculture.

Precision agriculture leverages modern technology to monitor, measure, and respond to variability in crops, soil, and environmental conditions within agricultural fields. This chapter highlights technologies such as sensors, drones, and hyperspectral imaging used in precision agriculture. Predictive analysis plays a crucial role in precision agriculture by utilizing advanced data analytics, ML models, and statistical techniques to forecast crop performance, soil health, and farming outcomes. Following the discussion on tools used in precision agriculture, the chapter explores predictive analysis techniques, including regression analysis, time-series analysis,

DOI: 10.1201/9781003589730-2

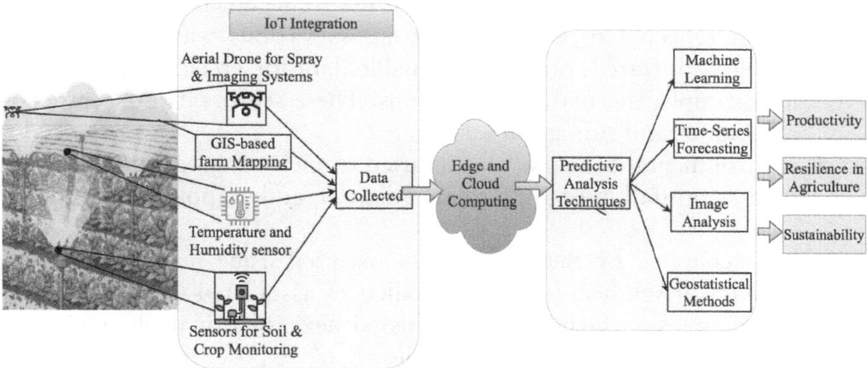

Figure 2.1 Predictive analysis in agriculture using IoT, edge and cloud computing.

ML, neural networks, and image analysis methods. The primary applications of precision agriculture emphasized in this chapter are crop yield prediction and optimization, stress detection, and soil mapping. The chapter is organized as follows: Section 2.1 introduces predictive analysis; Section 2.2 focuses on technologies used in precision agriculture; Section 2.3 discusses predictive analysis techniques in precision farming; Section 2.4 addresses challenges and future directions; and Section 2.5 concludes the chapter.

2.2 TECHNOLOGIES ENABLING PRECISION AGRICULTURE

2.2.1 Crop and soil sensors

Crop and soil sensors are crucial in smart agriculture. These sensors collect real-time data on soil conditions, crop health, and environmental factors, allowing farmers to make informed, data-driven decisions that result in improved agricultural outcomes. The popular types of sensors used in precision agriculture are as follows:

1. Electrochemical Sensors: These sensors are used to monitor the nutrient health of crops and soils, for example, ion-selective electrodes (ISE) for nitrate, phosphate, and potassium detection, pH sensors, salinity sensors, and heavy metal sensors.
2. Mechanical Sensors: Mechanical sensors measure physical properties like changes in wind velocities, pressure forces, accelerometers to track movement irregularities, etc.
3. Dielectric Sensors: Dielectric sensors measure soil moisture which is vital for plant growth. These sensors are also used for canopy monitoring.

4. Optical Sensors: Optical sensors use reflectance property. They detect reflected or emitted light across various wavelengths, from visible to infrared, providing valuable data for precision agriculture, for example, thermal vision systems. These sensors can measure soil reflectance and sunlight levels.
5. Environmental Sensors: Sensors used to measure weather conditions such as rainfall, humidity, solar radiation, and evapotranspiration.

The data generated by these sensors is analyzed using predictive models to make data-driven decisions. The challenges associated with using these sensors for precision farming are discussed next. Sensor deployment and maintenance are challenging as the sensors are exposed to variable environmental conditions. Power supply to sensors is another challenge in fields. Over time, sensors may lose their accuracy due to exposed stress. Hence, regular calibration of sensors is needed, requiring additional human efforts. Sometimes, dense deployment of sensors is needed to overcome the noise in data collected due to high spatial and temporal variability in fields, increasing deployment cost. Apart from the challenges associated with IoT network implementation and maintenance, integrating the diverse network at the edge is also challenging. The foremost challenge is the heterogeneity in data generated by different sensors. The data varies in unity, quantity, format, and type. The data generated is big and experiences overlapping, increasing the complexity of data management. Most importantly, the response time of an IoT device to the data collected depends on the network conditions between the IoT device and the edge device. Lastly, utilizing the benefits of IoT-based precision agriculture requires training and expertise.

2.2.2 Aerial drones

Aerial drones are unmanned aerial vehicles (UAVs) equipped with cameras, sensors, and Global Positioning System (GPS) technology for generating precise data. Primary usages of UAVs in agriculture are spraying fertilizers, generating high-resolution maps, and monitoring crop health. Camera sensors, fixed on drones, generate image data of the farmland, which is processed using image analysis methods at the edge or cloud devices. Thermal imaging cameras on drones are helpful in identifying wet and dry soil areas and finding applications in automated irrigation. Hyperspectral cameras on drones are more valuable than thermal cameras. Thermal cameras report information on temperature variation only. At the same time, hyperspectral cameras capture detailed spectral bands, providing detailed information on the chemical and physical properties of crops and soil. Figure 2.2 shows the images captured using spectral cameras and thermal cameras positioned on a drone. Apart from field monitoring, advanced drones are also used for seed plantations. Fixed-wing drones and multi-rotor drones are the

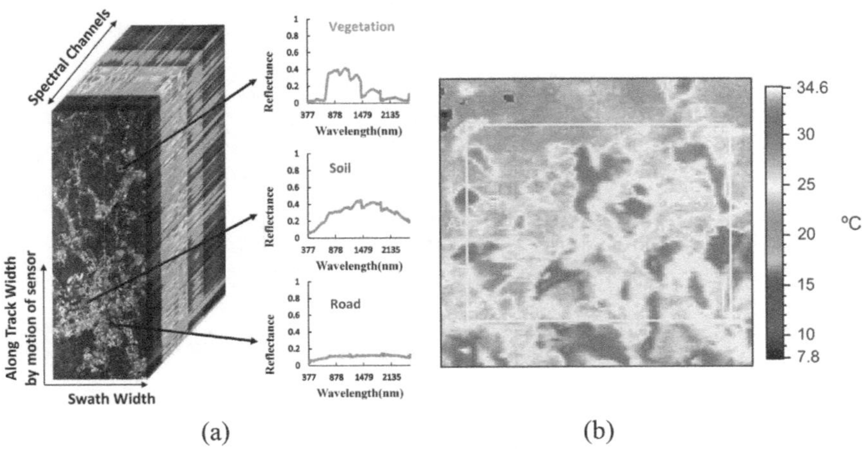

Figure 2.2 Images generated using hyperspectral (a) and thermal (b) cameras (on left [2] and right side [3], respectively).

most popular drones used in agriculture. Fixed-wing drones are helpful for field mapping and large-scale surveys due to their longer flight durations. Multi-rotor drones such as quadcopters, hexacopters, or octocopters are capable of vertical flights and are used for precise spraying of fertilizers and pesticides.

2.2.3 Farm mapping systems

The mapping of farmland in the new form of agriculture is an essential tool in modern agriculture, integrating geospatial technologies, sensor data, and the power of predictive models to generate very detailed maps of the farmlands. Farm mapping enables farmers to map their property, stating boundaries, fences, buildings, and water pipes placed on the geographical land. These systems provide a better understanding of farm planning as they enable precision farming. Figure 2.3 shows the major components of a farm mapping system and are discussed as follows:

1. Data Collection Components: consisting of soil sensors, crop sensors, drones, satellite weather stations, and GPS devices. GPS devices are helpful in providing the precise location of sensors' position and field boundaries.
2. Data Transmission Components: consisting of IoT gateway devices that collect data from devices at the acquisition layer and use wireless technologies such as Wi-Fi, LoRa, and Zigbee to communicate data using communication networks to processing units.

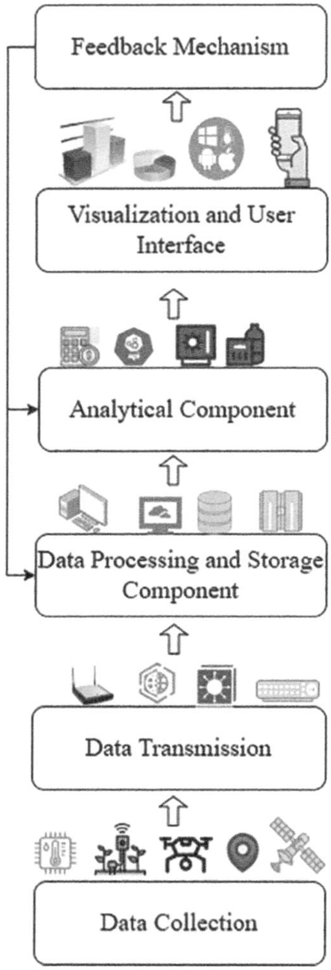

Figure 2.3 Components of a farm mapping system.

3. Data Processing and Storage Components: consisting of edge devices for fast real-time processing near farms and clouds for back-end high-processing and storage purposes.
4. Analytical Components: comprise a platform that supports spatial data analysis, predictive modeling, interpolations, image analysis, and statistical tools.
5. Visualization and User Interface Components: such as mobile applications and dashboards to visualize data in the form of maps, charts, and graphs.
6. Feedback Mechanisms: helpful in refining predictive models and improving map accuracy, alongside enabling manual adjustments from farmers.

2.2.4 GIS-based mapping for soil and crop monitoring

Geographic Information Systems (GISs) have emerged as transformative tools in agricultural research, enabling the precise monitoring and management of soil and crop conditions [4]. By integrating spatial data from diverse sources such as satellites, drones, IoT sensors, and weather stations, GIS facilitates the creation of high-resolution thematic maps that provide insights into soil fertility, moisture levels, and crop health [5,6]. Furthermore, GIS-based temporal analyses have enabled researchers to study long-term changes in soil quality and crop productivity under varying climatic and management conditions. GIS technology helps in mapping areas vulnerable to soil erosion and prediction of soil erosion patterns. For example, Pham et al. [7] combined the spatial data generated by GIS technology with the universal soil loss equation for reducing soil erosion in Vietnam.

Role in Spatial Variability Management: Research has demonstrated the utility of GIS in identifying spatial variability within fields, leading to the delineation of management zones for targeted interventions like site-specific fertilization, irrigation, and pest control [8]. Advanced spatial analysis techniques, such as kriging and Inverse Distance Weighting (IDW) [9,10], are frequently employed to interpolate missing data points and generate predictive models. The technologies mentioned above are integrated with edge and cloud computing to enable fast response and high processing, respectively. Table 2.1 summarizes using different technologies discussed above to enhance precision agriculture practices. 'Objective' column in Table 2.1 highlights the main objective of the work; 'Technology' contains the use of sensors, artificial intelligence (AI), ML, drones, GPS, and cloud computing; 'Environment' discusses the place considered by the work; 'Performance metrics' mentions the matrices used to evaluate the performance of their proposed models followed by the discussion of the results in the column 'Results.'

2.3 PREDICTIVE ANALYSIS TECHNIQUES

2.3.1 Regression analysis

Regression analysis is a statistical method that uses known independent variables to predict the behavior of an unknown dependent variable. For example, rainfall, temperature, and soil nutrients (independent variables) can be used to predict crop yield outcomes (dependent variable) [26]. This approach relies on historical data for both dependent and independent variables to develop a function that models the relationship and predicts outcomes accurately. The function can be a linear function or a nonlinear function. Linear functions have low computation complexity. However, nonlinear functions model the behavior more accurately and provide better

Table 2.1 Existing contributions of researchers

Ref. and year	*Objective*	*Technology*	*Environment*	*Performance metrics*	*Results*
Gulaiya et al. [11]	To enhance farming efficiency using smart technologies	Artificial intelligence (AI) and Internet of Things (IoT)	Agricultural fields in India	Crop yield, resource utilization	Improved accuracy and profitability in modern agriculture
Garg et al. [12]	To develop a multimodal system for precision agriculture	IoT and machine learning (ML)	Controlled agricultural settings	Crop damage prediction accuracy	Effective crop damage prediction and disease detection
Sethuramalingam and Arumuga Perumal [13]	To implement IoT in traditional agriculture	IoT, sensors, cloud computing	Traditional farmlands	Resource optimization, crop monitoring	Enhanced productivity and profit through precision agriculture
Yadav and Sidana [14]	To analyze the use of advanced technologies in precision agriculture	Drones, sensors, GPS	Various agricultural environments	Resource management efficiency, yield prediction accuracy	Optimized resource management and improved crop monitoring
Maddikunta et al. [15]	To explore unmanned aerial vehicle (UAV) applications in smart agriculture	UAVs	Smart farming environments	Application efficiency, operational challenges	Identified future applications and challenges of UAVs in agriculture
Shukla et al. [16]	To explore advancements in sensor-based technologies for precision agriculture	Sensors, IoT, ML	Diverse agricultural settings	Interoperability, data analytics effectiveness	Highlighted the potential of sensor technologies for sustainable agriculture

(*Continued*)

Table 2.1 (Continued) Existing contributions of researchers

Ref. and year	*Objective*	*Technology*	*Environment*	*Performance metrics*	*Results*
Singh [17]	To review advances in precision agriculture technologies	Remote sensing, automation, AI	Crop production systems	Yield improvement, input reduction	Potential for increased yields and profitability while reducing inputs
Manjunath [18]	To leverage AI in real-time crop monitoring	AI	Experimental farms	Crop health monitoring accuracy	Improved real-time crop health management
Rao [19]	To integrate cloud computing in agricultural decision-making	Cloud computing, IoT	Mixed farming systems	Data integration and decision-making efficiency	Reduced delays in agricultural decisions
Brown [20]	To utilize drones for pesticide application	Drones, remote sensing	Open agricultural fields	Pesticide coverage efficiency	Improved application accuracy and reduced waste
Lee [21]	To automate precision farming tasks using robotics	Robotics, sensors, AI	Controlled-environment agriculture	Task execution accuracy, resource use	Enhanced task automation and resource savings
Johnson [22]	To secure agricultural data using blockchain	Blockchain, IoT	Diverse agricultural systems	Data security, traceability	Improved trust and transparency in data handling
Thomas [23]	To implement ML for yield prediction	ML	Multiple cropping systems	Prediction accuracy, model efficiency	Higher yield prediction accuracy achieved
Anderson [24]	To develop smart irrigation systems	IoT, sensors, cloud platforms	Water-scarce regions	Water use efficiency	Reduced water consumption by 30%
Smith [25]	To design autonomous agricultural vehicles	Autonomous vehicles, GPS	Large-scale farms	Navigation efficiency, labor cost reduction	Enhanced efficiency in large-scale farming

accuracy than linear functions. The authors in [27] combined the powers of ML with a multiple linear regression (MLR) model to provide an interaction regression model for the yield prediction of corn and soybean crops with state-of-the-art accuracy.

2.3.2 Time-series analysis

Time-series analysis uses data indexed in time to predict future trends, optimize resource allocation, and improve decision-making. Time-series analysis uses Autoregressive Integrated Moving Average (ARIMA) and Long Short-Term Memory (LSTM) to predict future points based on historical trends. ARIMA uses autoregression (AR in ARIMA) to predict current and future values based on past values. ARIMA model requires stationary data, which means the statistical properties of the series, such as mean and autocorrelation, do not change over time. Agriculture data is non-stationary and is thus first converted to stationary series using methods like first-order differencing and second-order differencing. Addressing the non-stationarity in the data series constitutes the integrated component of ARIMA (I in ARIMA). Lastly, the moving average (MA in ARIMA) models the dependency of current values on past error terms. The ARIMA model is applied in agriculture for crop production forecasting using climate data such as rainfall and temperature [28]. The process includes four stages: identification, where the data stationarity and normality are ensured; estimation, which uses maximum likelihood methods for the calculation of the parameters; diagnostic checks, which verify residual uncorrelation, constant variance, and normal distribution through autocorrelation function and partial autocorrelation function analysis; and forecasting. The normality of residuals is checked graphically, and if the diagnostics pass, the model is quite effective in predicting agricultural parameters like rainfall and temperature from historical datasets, for example, 1967–2016.

2.3.3 ML models

ML algorithms in precision agriculture are primarily used for pest and weed identification, crop yield prediction, disease detection, and classification of different types of crops. Support vector machines (SVMs) can handle complex and nonlinear classification tasks efficiently. Therefore, SVM is well explored in precision agriculture for the accurate and timely detection of pests and weeds [29,30]. Random forests (RFs) are a popular choice where the dataset is large, high dimensional, and contains a mix of numerical and categorical data types. RF is a collection (forest) of multiple decision trees. The output generated by an RF is an average prediction of all the decision trees. Gradient Boosting (GB) ML technique, for example, XGBoost, builds trees sequentially, where each tree is an improved version of the previous tree. Du et al. [31] used GB model to design corn seeding rules based on soil fertility to improve the yield prediction pattern. The results show improvement over the yield prediction generated using RF model.

2.3.4 Image analysis for stress detection

Image analysis for stress detection in crops involves using advanced imaging techniques and algorithms to monitor plant health. Figure 2.4 shows images from the dataset [32] along with the class of stress mentioned. Plant stress can be either due to biotic stress such as insects, fungus, and bacteria or abiotic stress like environmental conditions as mentioned earlier. Stress factors like water scarcity, nutrient deficiency, disease, and pest infestation manifest as changes in color, texture, or temperature, which can be captured using imaging tools. Key technologies include:

- RGB Imaging: Detects visible stress signs like discoloration as shown in Figure 2.4.
- Thermal Imaging: Identifies water stress by detecting temperature variations.
- Hyperspectral Imaging: Analyzes spectral signatures to pinpoint stress at molecular levels.
- AI and ML: Processes and interprets image data for early stress detection and actionable insights.

As an example, a study in [33] involved using non-invasive thermal-RGB imagery and computer vision to diagnose water stress conditions in winter wheat crops to reduce yield loss associated with sub-optimal irrigation scheduling. Field experiments were conducted across 2 years over plots irrigated with flood or sprinkler-based systems at differing water levels from

Figure 2.4 Visible impact of plant stress on leaves [32].

100%, 75%, 50%, and 25% of the crop evapotranspiration values. Data was collected from 3200 images at different growth stages, and canopy temperature, relative water content, soil moisture, and relative humidity measurements were also taken. The results showed that the higher irrigation levels resulted in lower canopy temperatures and higher soil moisture and water content. Deep learning models (ResNet50, AlexNet, etc.) were used to process the imagery, and ResNet50 achieved a 96.9% accuracy with RGB and 98.4% with thermal imagery. DL-LSTM, in function-approximation models, was able to achieve 96.7% accuracy. The study concludes that thermal-RGB imagery and computer vision techniques work well together to manage crop water stress. Weeds and pests have always been and remain significant agri-environmental problems that are severely reducing the yields, quality, and long-term success of crops. Management of the problem in a smart way is the main key of precision agriculture. Image segmentation methods utilize imagery data to highlight or identify weeds or pests present in a cultivated area. Moazzam et al. [34] used drones to capture images of crop fields and identify weeds in the images. The authors designed a semantic segmentation algorithm that classifies vegetation in an image as a weed or crop at pixel levels. The other method is plant-based classification in which a bounding box around a plant is labeled as a weed or crop. Moazzam et al. in their work [34] highlighted that plant-based classification results in a localization problem and pixel-level classification generates improved results. Liu et al. [35] highlighted that the performance of ML-based segmentation models depends on the classes present in the training dataset, and it is not possible to add all weed classes in a dataset. Therefore, they designed a semantic segmentation method in which they designed a mask for corn plants, and every green plant pixel identified outside the mask is categorized as a weed.

2.3.5 Neural networks

Neural networks now play a transformative role in predictive analysis for agriculture by tapping into the complex ability to process complex, nonlinear relationships within data. Some applications include crop yield forecasting using environmental and soil parameters, pest and disease prediction using models like Recurrent Neural Networks (RNNs) and LSTMs, and analysis of drone imagery via convolution neural networks (CNNs) for identifying crop stress or nutrient deficiencies. These models provide high precision, scalability, and automation. The biggest challenge in the implementation of precision agriculture remains the need for quality datasets and computational resources. Figure 2.5 shows an example of using a neural network in a classification problem. Nodes (also called neurons) in the input layer of the network contain environmental features. The hidden layer assigns optimal weights to different parameters based on the training data. The final model is used to classify the crop variety that will provide the maximum

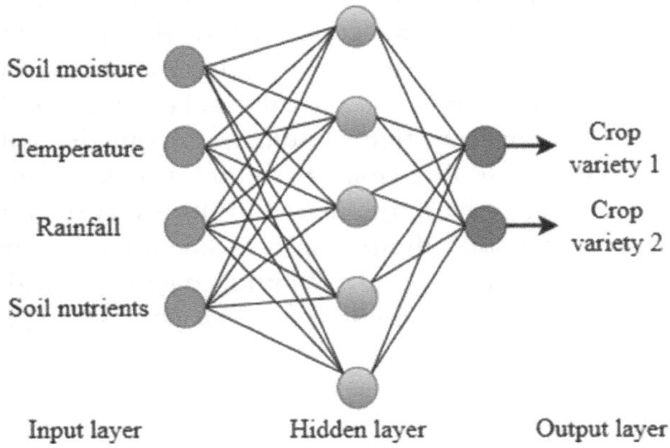

Figure 2.5 Simple neural network for agriculture classification problem.

yield given the weather specifications. Similarly, neural networks are used for regression tasks in which the output layer contains only one neuron with a continuous value, e.g., crop yield prediction. A simple neural network can be converted to deep neural networks by increasing the number of hidden layers, resulting in more accurate but complex networks. RNNs and LSTMs (advanced LSTMs) contain a recurrent unit that stores information about the previous inputs. The network comprises cycles allowing information to persist across time steps. This memory-like capability enables RNNs to analyze sequential data, such as time-series crop growth, weather patterns, or irrigation schedules, in agriculture. Researchers in [36] used RNNs to diagnose crop pests and diseases early, ensuring timely sprays and better crop management.

The study in [37] explores using MLR and artificial neural network (ANN) models to predict the yield of very early potato cultivars before harvest. Researchers compared the performance of both models, incorporating variables like soil properties, weather conditions, and cultivation practices. The findings demonstrated that ANNs outperformed MLR in prediction accuracy, highlighting the potential of ML in precision agriculture. The work emphasizes the importance of advanced predictive tools for optimizing crop management and supporting agricultural decision-making.

2.3.6 Spatial data analysis

Spatial data analysis involves collecting data from a certain region, analyzing it, and using the collected spatial information to improve agriculture practices. Geospatial and statistical tools used for agricultural projects have been optimized, which has resulted in improved farming practices, resource efficiency, and crop yield. With the aid of precision farming and sustainable land management, farmers and researchers can benefit from the data on the spatial distributions and interconnections of agricultural entities so as to be able to make the necessary decisions. Soil salinity is one of several spatial features responsible for deciding the agricultural land available for farming. It refers to the high concentration of soluble salts in the soil that negatively affects plant growth. Precision agriculture includes using technology and data analysis to map the soil salinity and suggest agriculture routines. Wang et al. designed a soil salinity risk assessment method using remote sensing and GIS [38]. Kingsley et al. [39] utilized geostatistical methods to map and predict the spatial variability of soil properties like pH, phosphorus, soil organic carbon (SOC), clay content, and nitrogen.

2.3.7 Statistical tools

Analysis of Variance (ANOVA) is a statistical technique used to determine whether there is a statistically significant difference between the means of independent, unrelated groups. Khanna and Kaur [40] applied ANOVA to questionnaire response data collected from 342 farmers in Punjab, India. The questionnaire focused on the challenges faced by these farmers in adopting precision farming. The primary constraints identified were a lack of awareness, financial limitations, and limited technical knowledge. The authors divided the farmers into two groups based on their prior knowledge of agricultural sensors and their usability. The one-way ANOVA revealed that the two groups differed significantly. Additionally, an independent-sample *t*-test showed that the group with greater knowledge exhibited a stronger inclination toward adopting modern agricultural practices compared to the less knowledgeable group. Another study by Balogh et al. [41] examines the socio-economic aspects of precision farming in Hungarian agriculture. The collected data was divided into three clusters using two-step clustering. The correlation among clusters was analyzed using ANOVA and Chi-square tests. In addition to statistical tools for survey analysis, dimensionality reduction techniques, such as principal component analysis (PCA), are also available.

2.4 CHALLENGES AND FUTURE DIRECTIONS

Predictive analysis can greatly contribute to precision agriculture in enhancing farm mapping, crop stress detection, and the decision-making processes involved. However, challenges continue to arise and require new innovative methods for full implementation.

2.4.1 Challenges

1. Data Collection and Quality: Data is the foremost requirement of statistical and ML algorithms used in precision agriculture. However,

high-resolution spatial and temporal data is rare in developing regions, with less availability. Simultaneously, the data generated is heterogeneous and requires expert knowledge in both integration and analysis.
2. Technological Challenges: Deploying crop and soil sensors and maintaining broadband connectivity is challenging in villages, creating an infrastructure barrier. Sensors, drones, and other precision agriculture tools are expensive and thus unaffordable to small-scale farmers to adopt them.
3. Cost Concerns: Cost of precision agriculture equipment is too high. Even after bearing the deployment cost, farmers need money for training and maintenance purposes.
4. Adoption and Awareness: Limited knowledge and skills among farmers to utilize advanced precision agriculture tools effectively.

2.4.2 Future directions

1. Advanced Data Integration: The current on-field agriculture system lags in data integration methodology, integrating data from satellites, drones, sensors, and GPS. Statistical models will integrate time-series data with spatial information to track changes and trends in soil and crop conditions. A combined information from multiple sources will help in improving accuracy of target-specific applications, for example, smart irrigation systems, livestock monitoring, and supply chain optimization.
2. Micro Zone Analysis in a Farm Mapping System: Statistical tools will help identify small-scale variability within fields for more precise resource application.
3. Risk Assessment: Statistical simulations and probabilistic models will identify potential risks, such as pest outbreaks or water stress, and suggest preventive actions.
4. Dynamic Mapping for Real-Time Decision Support: Statistical computations performed on IoT devices in the field will provide immediate insights without relying on cloud processing.
5. Statistical Methods for Advanced Visualization: Statistical methods will enable the creation of 3D maps showing sub-surface data, and 4D maps will visualize temporal changes in farm conditions.
6. Targeted Solutions for Smallholder Farmers: There is a scope of designing affordable tools. Create low-cost sensors and analytics platforms for resource-poor farmers. Community-based models can be designed to implement cooperative models to share precision agriculture tools and knowledge.
7. Policy and Education: Encourage public-private partnerships and subsidies to embrace precision agriculture. Design farm-centric training programs to enhance the adoption of technology.

2.5 CONCLUSION

Predictive analytics has become an indispensable tool in modern agriculture to bring innovation to increasing production, sustainability, and resilience. Advances in analytical methods, including regression analysis, time-series forecasting, ML, neural network techniques, and geospatial analytics, permit farmers and researchers to extract meaningful insights from crop and soil sensor data. In this context, the integration with IoT and edge computing enables real-time decision-making, ensuring the prompt resolution of any challenge and resource optimization. The application of these techniques is projected to range from early disease detection and yield prediction to precision irrigation and fertilization, thereby promoting sustainable farming practices addressing those global challenges posed by climate change, resource scarcity, and food security. As technology evolves, this synergy promises predictive models in the future of agriculture that will combine with farming systems from future to initiate data-driven decision-making to meet an ever-increasing world-at-a-high-speed demand. Consequently, we have a chapter that discusses how predictive analytics has the potential to redefine how agriculture is conducted and establish smarter, more sustainable agricultural landscapes.

REFERENCES

1. N. Singh and M. Adhikari, "Edge-centric collaborative federated learning for irrigation management of paddy fields using agriculture sensor data processing," *IEEE Sensors Letters*, vol. 8, p. 6009104, 2024.
2. S. S. Jha, Multi-platform Hyperspectral Target Detection and Modelling in Dynamis Atmospheric Conditions. PhD thesis, Indian Institute of Space Science and Technology, 2021.
3. D. Budiman, E. Asri, and Z. R. Aidha, "The smart agriculture based on reconstructed thermal image," in *2022 2nd International Conference on Intelligent Technologies (CONIT)*, Hubli, India, pp. 1–6, IEEE, 2022.
4. A. Raihan, "A systematic review of geographic information systems (GIS) in agriculture for evidence-based decision making and sustainability," *Global Sustainability Research*, vol. 3, no. 1, pp. 1–24, 2024.
5. M. A. Abdellatif, A. A. El Baroudy, M. Arshad, E. K. Mahmoud, A. M. Saleh, F. S. Moghanm, K. H. Shaltout, E. M. Eid, and M. S. Shokr, "A GIS-based approach for the quantitative assessment of soil quality and sustainable agriculture," *Sustainability*, vol. 13, no. 23, p. 13438, 2021.
6. I. Dunaieva, W. Mirschel, V. Popovych, V. Pashtetsky, E. Golovastova, V. Vecherkov, A. Melnichuk, V. Terleev, A. Nikonorov, R. Ginevsky, et al., "GIS services for agriculture monitoring and forecasting: Development concept," in *International Scientific Conference Energy Management of Municipal Facilities and Sustainable Energy Technologies EMMFT 2018*, pp. 236–246, Springer, 2018.

7. T. G. Pham, J. Degener, and M. Kappas, "Integrated universal soil loss equation (USLE) and geographical information system (GIS) for soil erosion estimation in a sap basin: Central Vietnam," *International Soil and Water Conservation Research*, vol. 6, no. 2, pp. 99–110, 2018.
8. S. H. Rano, M. Afroz, and M. M. Rahman, "Application of GIS on monitoring agricultural insect pests: A review," *Reviews in Food and Agriculture*, vol. 3, no. 1, pp. 19–23, 2022.
9. D. Zimmerman, C. Pavlik, A. Ruggles, and M. P. Armstrong, "An experimental comparison of ordinary and universal kriging and inverse distance weighting," *Mathematical Geology*, vol. 31, pp. 375–390, 1999.
10. T. Mueller, N. Pusuluri, K. Mathias, P. Cornelius, R. Barnhisel, and S. Shearer, "Map quality for ordinary kriging and inverse distance weighted interpolation," *Soil Science Society of America Journal*, vol. 68, no. 6, pp. 2042–2047, 2004.
11. S. Gulaiya, A. Sharma, S. Singh, R. Kumar, D. Chaudhary, K. Yadav, M. Kumar, and R. Kumar, "Precision agriculture: Transforming farming efficiency with cutting-edge smart technologies: A comprehensive review," *Journal of Experimental Agriculture International*, vol. 46, no. 9, pp. 1126–1138, 2024.
12. S. Garg, P. Pundir, H. Jindal, H. Saini, and S. Garg, "Towards a multimodal system for precision agriculture using IoT and machine learning," in *2021 12th International Conference on Computing Communication and Networking Technologies (ICCCNT)*, Kharagpur, India, pp. 1–7, IEEE, 2021. https://doi.org/10.48550/arXiv.2107.04895.
13. S. Sethuramalingam and S. Arumuga Perumal, "IoT for precision agriculture," *International Journal of Advance Research, Ideas and Innovations in Technology*, vol. 8, pp. 409–415, 2023.
14. N. Yadav and N. Sidana, "Precision agriculture technologies: analysing the use of advanced technologies, such as drones, sensors, and GPS, in precision agriculture for optimizing resource management, crop monitoring, and yield prediction," *Journal of Advanced Zoology*, vol. 44, no. S5, pp. 255–263, 2023.
15. P. K. R. Maddikunta, S. Hakak, M. Alazab, S. Bhattacharya, T. R. Gadekallu, W. Z. Khan, and Q.-V. Pham, "Unmanned aerial vehicles in smart agriculture: Applications, requirements and challenges," arXiv preprint, vol. arXiv:2007.12874, 2020.
16. B. K. Shukla, N. Maurya, and M. Sharma, "Advancements in sensor-based agriculture," *Engineering Proceedings*, vol. 58, p. 22, 2023.
17. V. Singh, "Advances in precision agriculture technologies for sustainable crop production," *Journal of Scientific Research and Reports*, vol. 30, pp. 61–71, 2024.
18. R. Manjunath, "AI for crop monitoring," *Experimental Agriculture Journal*, vol. 57, pp. 123–133, 2022.
19. K. Rao, "Cloud integration in agriculture," *Agricultural Systems and Technologies*, vol. 45, pp. 101–110, 2021.
20. P. Brown, "Drone-based pesticide application," *Precision Agriculture Journal*, vol. 12, pp. 231–242, 2020.

21. T. Lee, "Robotics for farming tasks," *Automation in Agriculture*, vol. 34, pp. 45–56, 2023.
22. A. Johnson, "Blockchain in agriculture," *Sustainable Agriculture Technologies*, vol. 8, pp. 112–120, 2022.
23. G. Thomas, "Machine learning for yield prediction," *Computational Agriculture and Data Science*, vol. 22, pp. 77–89, 2020.
24. J. Anderson, "Smart irrigation systems," *Water Management in Agriculture*, vol. 19, pp. 310–320, 2021.
25. D. Smith, "Autonomous agricultural vehicles," *Journal of Agricultural Robotics*, vol. 29, pp. 67–75, 2023.
26. S. Khaki, H. Pham, and L. Wang, "Simultaneous corn and soybean yield prediction from remote sensing data using deep transfer learning," *Scientific Reports*, vol. 11, no. 1, p. 11132, 2021.
27. J. Ansarifar, L. Wang, and S. V. Archontoulis, "An interaction regression model for crop yield prediction," *Scientific Reports*, vol. 11, no. 1, pp. 1–14, 2021.
28. B. Praveen and P. Sharma, "Climate variability and its impacts on agriculture production and future prediction using autoregressive integrated moving average method (ARIMA)," *Journal of Public Affairs*, vol. 20, no. 2, p. e2016, 2020.
29. F. Ahmed, H. A. Al-Mamun, A. H. Bari, E. Hossain, and P. Kwan, "Classification of crops and weeds from digital images: A support vector machine approach," *Crop Protection*, vol. 40, pp. 98–104, 2012.
30. Y. Wu, Y. He, and Y. Wang, "Multi-class weed recognition using hybrid CNN-SVM classifier," *Sensors*, vol. 23, no. 16, p. 7153, 2023.
31. Z. Du, L. Yang, D. Zhang, T. Cui, X. He, T. Xiao, C. Xie, and H. Li, "Corn variable-rate seeding decision based on gradient boosting decision tree model," *Computers and Electronics in Agriculture*, vol. 198, p. 107025, 2022.
32. N. A. Orka, M. N. Uddin, F. M. Toushique, and M. S. Hossain, "Olid I: An open leaf image dataset for plant stress recognition," *Frontiers in Plant Science*, vol. 14, p. 1251888, 2023.
33. N. S. Chandel, Y. A. Rajwade, K. Dubey, A. K. Chandel, A. Subeesh, and M. K. Tiwari, "Water stress identification of winter wheat crop with state-of-the-art AI techniques and high resolution thermal-RGB imagery," *Plants*, vol. 11, no. 23, p. 3344, 2022.
34. S. I. Moazzam, U. S. Khan, W. S. Qureshi, T. Nawaz, and F. Kunwar, "Towards automated weed detection through two-stage semantic segmentation of tobacco and weed pixels in aerial imagery," *Smart Agricultural Technology*, vol. 4, p. 100142, 2023.
35. T. Liu, X. Jin, K. Han, F. He, J. Wang, X. Chen, X. Kong, and J. Yu, "Semantic segmentation for weed detection in corn," *Pest Management Science*, vol. 81, pp. 1512–1528, 2024.
36. V. K. Patel, K. Abhishek, and S. Selvarajan, "Optimized recurrent neural network-based early diagnosis of crop pest and diseases in agriculture," *Discover Computing*, vol. 27, no. 1, p. 43, 2024.
37. M. Piekutowska, G. Niedbała, T. Piskier, T. Lenartowicz, K. Pilarski, T. Wojciechowski, A. A. Pilarska, and A. Czechowska-Kosacka, "The application of multiple linear regression and artificial neural network models for yield prediction of very early potato cultivars before harvest," *Agronomy*, vol. 11, no. 5, p. 885, 2021.

38. Z. Wang, F. Zhang, X. Zhang, N. W. Chan, H.-T. Kung, X. Zhou, and Y. Wang, "Quantitative evaluation of spatial and temporal variation of soil salinization risk using GIS-based geostatistical method," *Remote Sensing*, vol. 12, no. 15, p. 2405, 2020.
39. J. Kingsley, S. O. Lawani, A. O. Esther, K. M. Ndiye, O. J. Sunday, and V. Penížek, "Predictive mapping of soil properties for precision agriculture using geographic information system (GIS) based geostatistics models," *Modern Applied Science*, vol. 13, no. 10, pp. 60–77, 2019.
40. A. Khanna and S. Kaur, "An empirical analysis on adoption of precision agricultural techniques among farmers of Punjab for efficient land administration," *Land Use Policy*, vol. 126, p. 106533, 2023.
41. P. Balogh, A. Bai, I. Czibere, I. Kovách, L. Fodor, Á. Bujdos, D. Sulyok, Z. Gabnai, and Z. Birkner, "Economic and social barriers of precision farming in Hungary," *Agronomy*, vol. 11, no. 6, p. 1112, 2021.

Chapter 3

Predictive analysis in precision agriculture

Leveraging IoT for early detection of crop stresses and sustainable farming

Venkata Naga Rani Bandaru, Srikanth Cherukuvada, Shaik Mohammad Rafee, Idamakanti Kasireddy, and Pilli Suneetha

3.1 INTRODUCTION

3.1.1 Introduction of agriculture system

Precision farming, sometimes referred to as precision agriculture, is a technical advancement in the field of agriculture. It is a data-driven strategy that includes tracking, quantifying, and reacting to crop variability both within and between fields. Precision agriculture aims to maximize agricultural yields and reduce environmental impact while optimizing resource utilization. The idea has developed from conventional farming techniques, which usually administer uniform treatments over whole fields, to more focused approaches that modify inputs such as water, fertilizers, and pesticides according to the particular requirements of crops in various sections of the field. The emergence of GPS (Global Positioning System) technology in the early 1980s gave rise to precision agriculture as a field mapping technique that was previously unheard of in terms of accuracy. This innovation led to the development of site-specific agricultural techniques, which allowed for distinct management of different fields. Farmers could now measure crop output in real time during harvest thanks to the development of yield monitors in the 1990s. Farmers were able to produce yield maps of their fields using GPS and this data, which indicated whether areas need more or less input. In the 2000s, the usage of Geographic Information Systems (GIS) and remote sensing technology expanded the capabilities of precision farming. While remote sensing, which uses satellites and drones, provided a higher-up view of crop health and field conditions, GIS technologies allowed farmers to collect, store, process, and display geographical and temporal data. Precision agriculture has also come to rely heavily on variable rate technology (VRT), which enables machinery to dynamically modify input applications (seeds, fertilizers, and herbicides) in response to field data.

Some of the major issues confronting modern farming, such as the growing population's desire for food, the scarcity of arable land, and the negative

 DOI: 10.1201/9781003589730-3

environmental effects of intensive agricultural techniques, might be resolved by precision agriculture. Precision farming may assist in providing sustainable solutions to fulfill the world's expanding food demands by maximizing resource utilization and raising crop yields.

3.1.2 Importance of early crop stress detection

Precision agriculture depends on early diagnosis of crop stress since crop output and quality can be greatly impacted by its neglect. Conventional approaches are reactive rather than proactive as they frequently rely on time-consuming visual examinations. By enabling farmers to administer treatments customized to certain locations rather than treating the entire field consistently, this strategy promotes site-specific management. Farmers may successfully manage inputs and prevent possible losses by utilizing real-time data.

3.1.3 Role of IoT and predictive analytics in agriculture: a literature survey

Over the past 10 years, several studies have examined Internet of Things (IoT) and predictive analytics in agriculture, showing a dramatic change in farming practices. IoT-based climate-resilient farming systems allow farmers to identify stress factors like drought or floods early on, as Sharma et al. (2024a) show. In a similar vein, Kim and colleagues (2024) stress the use of IoT and machine learning to enhance agricultural output forecasts.

According to Srinivasan et al. (2023), data-driven insights are produced by combining big data analytics with IoT sensors to optimize resource consumption and enhance farming sustainability. With the use of IoT sensors that track soil moisture in real time, Prakash et al. (2023) concentrate on artificial intelligence (AI)-powered smart irrigation systems that enable water-efficient crop management.

Going back to prior research, Mudholkar et al. (2022) demonstrated how water use may be decreased while preserving optimal crop health with IoT and AI-driven smart irrigation systems. The influence of real-time monitoring on crop health is further demonstrated by Xie et al. (2022), who show how precision agricultural techniques may boost production by modifying inputs in response to real-time sensor data.

Hogeveen and Van der Voort (2021) demonstrate the usefulness of IoT in monitoring animal health and behavior in the context of livestock farming, whereas Yang et al. (2022) illustrate how the combination of cloud computing with IoT leads to more precise crop production estimates. The increasing interest in cloud-based predictive analytics, which enables farmers to improve output based on real-time data, is shown in this research.

Bendre et al. (2020) and Kumar et al. (2020) demonstrate how machine learning, in conjunction with drones and IoT sensors, aids in the monitoring of crop health and pests. According to their results, farmers may identify crop stress early using these devices, which enhances pest control and lowers the need for chemical treatments.

Patel et al. (2019) and Zhang et al. (2019) investigated the use of IoT in pest and disease detection as well, creating IoT-enabled systems that forecast agricultural illnesses and pest infestations. These devices allowed farmers to take action before the harm worsened by using environmental data to provide early warnings. The use of IoT in precision agriculture was pioneered by earlier research by Li et al. (2018), Jain et al. (2017), and Kamilaris et al. (2016), which optimized water efficiency and predicted crop yields using IoT-enabled sensors. Their results show that real-time field conditions combined with data-driven decision-making may greatly increase agricultural output while reducing resource waste.

The last 8 years' worth of research has demonstrated how IoT and predictive analytics are revolutionizing the agricultural sector. Along with making farming operations more efficient, these technologies have also contributed to developing long-term solutions for some of agriculture's most urgent problems, such as climate change, insect control, and water shortages (Table 3.1).

Table 3.1 Recent advances in agriculture

Author and Year	*Title*	*Methodology*
Sharma et al. 2024a	IoT and Predictive Analytics in Climate-Resilient Farming	IoT sensors, climate data analysis
Kim et al. 2024	Integrating Machine Learning into IoT-Based Precision Farming	Machine learning, IoT
Srinivasan et al. 2023	Big Data Analytics in IoT Agriculture	Big data, IoT, predictive models
Prakash et al. 2023	AI-Driven Smart Irrigation Using IoT	AI, IoT sensors, machine learning
Mudholkar et al. 2022	IoT and AI in Smart Irrigation Systems	AI, IoT sensors, machine learning
Xie et al. 2022	IoT-Based Monitoring for Crop Health	IoT sensors, data analytics
Van der Voort et al. 2021	Precision Livestock Farming Using IoT	IoT sensors, predictive algorithms
Yang et al. 2022	IoT and Cloud-Based Predictive Analytics for Crop Yield	IoT, cloud computing, data analytics
Bendre et al. 2020	IoT and Analytics for Crop Health Monitoring	Drones, IoT sensors, machine learning
Kumar et al. 2020	IoT-Enhanced Pest Monitoring in Agriculture	IoT, machine learning, pest detection
Patel et al. 2019	Pest Forecasting with IoT and Analytics	IoT sensors, environmental monitoring

(*Continued*)

Table 3.1 (Continued) Recent advances in agriculture

Author and Year	*Title*	*Methodology*
Zhang et al. 2019	Predictive Analytics for Disease Detection in Crops	Machine learning, environmental data
Li et al. 2018	IoT-Based Monitoring for Water Efficiency	IoT, water usage, predictive analytics
Jain et al. 2017	Predictive Models for Crop Yield Optimization	Machine learning, yield prediction
Kamilaris et al. 2016	IoT-Based Smart Farming	IoT, real-time monitoring

3.2 INTEGRATED IoT-DRIVEN PREDICTIVE AGRICULTURE SYSTEM ARCHITECTURE FOR SUSTAINABLE FARMING

To improve precision agriculture, the IoT-Driven Predictive Agriculture System (IPAS) combines sophisticated sensor technologies, reliable communication networks, data analytics, and automation.

3.2.1 Overview of IPAS

The IPAS framework comprises the following core components:

1. Sensors (S): A set of sensors $S = \{S_1, S_2, \ldots, S_n\}$ where each S_i represents a specific type of sensor (soil moisture pH).
2. Communication Networks (C): Utilized communications technologies $C = \{C_{\text{LoRa}}, C_{5G}\}$ for data transmission.
3. Edge Devices and Gateways (E): Edge computing devices.
4. Cloud Platforms (P): Cloud services $P\ (\text{AWS})$ for storage and advanced analytics.
5. Data Analytics Tools (A): Algorithm and models A developed using programming languages like Python.
6. Automation System (M): Mechanisms M such as automated irrigation that act based on processed data.

The main goal is to maximize crop yield Y while minimizing resource usage R:

$$\max_{S,C,E,P,A,M} (Y - \alpha R) \tag{3.1}$$

where α is a weighting factor representing the cost of resources (Figure 3.1).

3.2.2 Sensor integration

By integrating several sensors, farmers may receive real-time data that is vital for predictive analytics in agriculture. This data helps farmers make

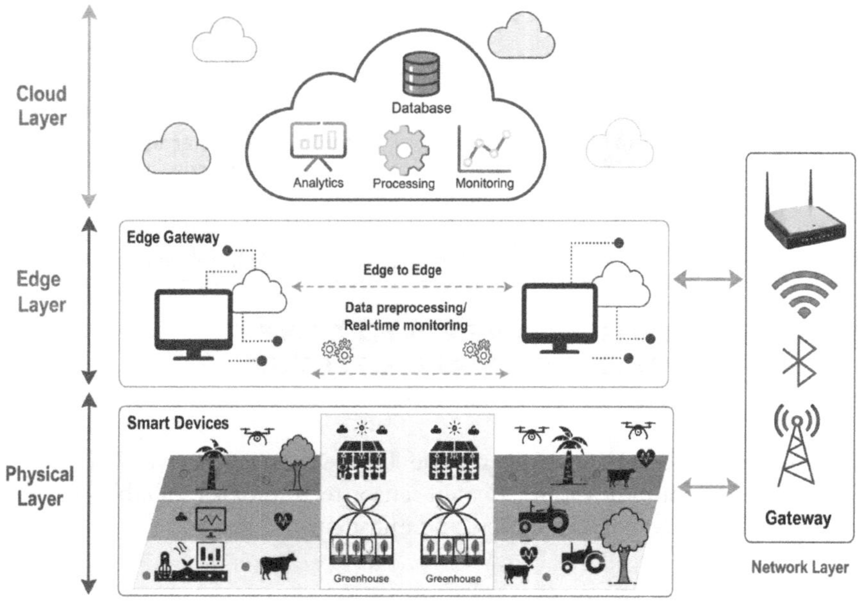

Figure 3.1 IPAS architecture: integrated IoT-driven predictive agriculture system.

educated decisions about crop health, resource management, and overall farm performance.

3.2.2.1 Soil moisture sensors

Utilizing the dielectric permittivity technique, irrigation depth (I_d), and volumetric water content (θ) in agriculture offers a complete way to manage soil moisture. When these factors are combined, farmers may increase crop health, optimize water use, and advance sustainable farming methods (Table 3.2).

3.2.2.2 pH sensors

3.2.2.2.1 Soil pH measurement

The pH of the soil is determined using the following formula:

$$\text{pH} = -\log_{10}\left[H^+\right] \tag{3.2}$$

where $\left[H^+\right]$ is the concentration of hydrogen ions in the soil solution. Knowing the pH of your soil is important because it may tell you how acidic or alkaline your soil is. The pH scale often runs from 0 (extremely acidic)

Table 3.2 Soil moisture parameters and calculations

Parameter	*Description*	*Formula*	*Explanation*
Volumetric Water Content (θ)	Measures the amount of water in the soil	$\theta = \frac{V_\omega}{V_t}$	V_ω:The volume of water in the soil. V_t :Total volume of the soil sample
Dielectric Permittivity Method	Uses the soil's dielectric constant to gauge moisture content	$\theta = a_0 + a_1\epsilon + a_2\epsilon^2$	a_1, a_2, a_3: Calibration coefficients vary by sensor and soil type. ϵ: The dielectric constant of soil shows its moisture content
Irrigation Depth (I_d)	Calculates the amount of water needed for irrigation.	$I_d = (\theta_{FC} - \theta_A) \times Z_r$	θ_{FC}: Field capacity, θ_A: moisture content. Z_r: root zone depth

to 14 (very alkaline), with 7 being neutral. The bioavailability of nutrients to plants and other biochemical processes in the soil is impacted by pH. For instance, a pH of less than 6.0 may increase the availability of hazardous metals, but a pH of more than 7.5 may cause deficits in some nutrients, especially micronutrients. If the hydrogen ion concentration in the soil solution is $\left[H^+\right] = 1.0 \times 10^{-5}$ mol/L, the pH would be calculated as:

$$\text{pH} = -\log_{10}\left(1.0 \times 10^{-5}\right) = 5.0$$

This pH indicated slightly acidic soil, which might require adjustments depending on the crop's nutrient requirements.

3.2.2.2.2 Effect on nutrient availability

The availability of nutrients in the soil can be modeled concerning pH using the equation:

$$N_a = f(\text{pH}) = \frac{1}{1 + e^{-k(\text{pH} - \text{pH}_0)}} \tag{3.3}$$

where N_a is the nutrient availability function, indicating the proportion of nutrients available for plant uptake, k is a rate constant that affects the steepness of the curve, showing how quickly nutrient availability changes with pH, and pH_0 is the optimal pH for nutrient uptake, typically varying by nutrient type and plant species. Nutrient availability is influenced by soil pH. For instance, at a pH significantly lower or higher than pH_0, nutrient availability may drop, signaling the need for amendments to optimize growth conditions.

At pH = 5.5:

$$N_a = f(\text{pH}) = \frac{1}{1+e^{-0.5(5.5-6.5)}} = \frac{1}{1+\ e^{0.5}} \approx 0.377$$

At pH = 6.5:

$$N_a = f(\text{pH}) = \frac{1}{1+e^{-0.5(6.5-6.5)}} = \frac{1}{1+\ e^{0}} = \frac{1}{2} = 0.5$$

At pH = 7.5:

$$N_a = f(\text{pH}) = \frac{1}{1+e^{-0.5(7.5-6.5)}} = \frac{1}{1+\ e^{-0.5}} \approx 0.622$$

According to these calculations, nutrient availability rises as pH gets closer to the ideal level of 6.5, emphasizing the need of pH monitoring and management for the best possible crop development.

3.2.2.3 Temperature sensor

3.2.2.3.1 Degree days calculation

A measure called Cumulative Growing Degree Days (GDD) is used to calculate the amount of heat buildup needed for crop development and growth. Based on daily temperature changes, it aids in the prediction of plant growth phases, from germination to maturity. This formula is used to determine GDD:

$$\text{GDD} = \sum_{i=1}^{n}\left(\frac{T_{\max,i}+T_{\min,i}}{2} - T_{\text{base}}\right) \tag{3.4}$$

where $T_{\max,i}$ is the daily maximum temperature on day i, $T_{\min,i}$ is the daily minimum temperature on day i, and T_{base} is the base temperature below which the crop growth is negligible (Table 3.3).

Day 1 GDD:

$$\text{GDD}_1 = \frac{25+15}{2} - 10 = \frac{40}{2} - 10 = 20 - 10 = 10$$

Table 3.3 Daily temperature records

Day	T_{max} (°C)	T_{min} (°C)
1	25	15
2	20	12
3	18	8

Day 2 GDD:

$$GDD_2 = \frac{20+12}{2} - 10 = \frac{32}{2} - 10 = 16 - 10 = 6$$

Day 3 GDD:

$$GDD_3 = \frac{18+8}{2} - 10 = \frac{26}{2} - 10 = 13 - 10 = 3$$

Total GDD for 3 days:

$$GDD_{total} = GDD_1 + GDD_2 + GDD_3 = 10 + 6 + 3 = 19$$

Over 3 days, the total GDD is 19. To ascertain the crop's developmental stage—when to anticipate blooming or harvest—this figure may be compared to the crop's heat needs.

3.2.2.4 Humidity sensor

3.2.2.4.1 Relative humidity calculation

A crucial metric in agriculture is relative humidity (RH), which expresses the proportion of air moisture to the maximum amount of moisture the air can contain at a certain temperature (saturation vapor pressure). It has a direct impact on transpiration and water absorption in plants. The following is the formula to find RH:

$$RH = \left(\frac{e}{e_s}\right) \times 100\% \tag{3.5}$$

where e is the actual vapor pressure, representing the current moisture content in the air and e_s is the saturation vapor pressure which is the maximum amount of moisture the air can hold at the current temperature.

3.2.2.4.2 Actual vapor pressure calculation

If the saturation vapor pressure and RH are known, the actual vapor pressure may be computed. The connection is provided by:

$$e = RH \times e_s \tag{3.6}$$

where RH is the relative humidity (as a decimal, not a percentage) and e_s is the saturation vapor pressure.

3.2.2.5 Nutrient sensor

3.2.2.5.1 Soil nutrient content estimation

Soil nutrient sensors provide the assessment of the concentration of particular nutrients in a given soil sample. Equation: This may be used to determine the nutrient content (C_N).

$$C_N = \frac{M_N}{V_s} \tag{3.7}$$

where M_N is the mass of the nutrient in the soil sample, typically measured in milligrams (mg) or grams (g), and V_s is the volume of the soil sample, measured in cubic centimeters (cm^3) or liters (L). By figuring out how much of a nutrient is in the soil, this formula aids in determining the fertility of the soil. For instance, the concentration of nitrogen (C_N) would be as follows if a soil sample contained 50 mg of nitrogen (M_N) in a 100 cm^3 soil sample (V_s):

$$C_N = \frac{50}{100} = 0.5 \text{ mg/cm}^3$$

3.2.2.5.2 Fertilizer recommendation model

The required amount of fertilizer (F_N) for a field can be determined by the following formula:

$$F_N = \left(N_{\text{target}} - C_N\right) \times A \times \text{Efficiency} \tag{3.8}$$

where N_{target} is the target nutrient level that is optimal for crop growth (measured in mg/cm^3 or g/m^3), C_N is the current nutrient level in the soil, A is the area of the field that needs fertilization (measured in square meters, m^3) and Efficiency is the percentage of the applied fertilizer that the crops will effectively utilize.

$$F_N = (1.0 - 0.5) \times 1000 \times 0.7 = 350 \frac{\text{mg}}{\text{m}^2}$$

For example, if the goal level is 1.0 mg/cm^3 (N_{target}) and the field's present nitrogen concentration is 0.5 mg/cm^3 (C_N), this is the amount of fertilizer needed for a 1000 m^2 field with 70% nutrient absorption efficiency.

3.2.2.6 Light sensors

3.2.2.6.1 Photosynthetically active radiation

The part of sunlight (in the 400–700 nm wavelength range) that plants can use for photosynthesis is known as photosynthetically active radiation or PAR. The formula for calculating PAR is:

$$\text{PAR} = \int_{\lambda_1}^{\lambda_2} E(\lambda)\,d\lambda \tag{3.9}$$

where (λ) Spectral irradiance is the intensity of light at a specific wavelength and λ_1 and λ_1. The wavelength range is typically between 400 and 700 nm.

3.2.2.6.2 Crop light utilization efficiency (ϵ)

Crop Light Utilization Efficiency (ϵ) describes how efficiently plants convert the available light (PAR) into biomass. It is calculated as:

$$\epsilon = \frac{\text{Biomass Accumulation}}{\text{PAR}} \tag{3.10}$$

where Biomass Accumulation is the increase in plant biomass (measured in g/m^2 or kg/ha) over a specific period and PAR, photosynthetically active radiation, represents the amount of light energy available.

3.2.2.6.3 Pest detection algorithm

Initialize:

1: Set feature set F, weight vector ω, and bias b.

2: Input sensor data for pest detection (features representing environmental and plant conditions)

3: Define the maximum number of iterations max_iter (for optimization, if needed).

4: For each data sample i.

5: Compute the decision value $z_i = \omega^T F_i + b$, where ω^T is the transpose of the weight vector and F_i is the feature vector for the sample i.

6: Apply the sign function to classify the sample: $\hat{y}_i = \text{sign}(z_i)$

$$= \begin{cases} 1 \text{ if } z_i < 0 & (\text{pest detected}) \\ -1 \text{ if } z_i < 0 & (\text{no pest detected}) \end{cases}$$

7: If $\hat{y}_i = 1$, alert for pest presence, otherwise no action.

8: Repeat for all data samples.

End

This method analyzes the input characteristics to determine if a pest is present or not based on sensor data using the linear classifier $y = \text{sign}(\omega^T F + b)$. It facilitates the prompt delivery of pest management notifications.

3.2.3 Communication networks

IoT-driven predictive agricultural systems rely on effective communication to transfer data seamlessly between sensors, controllers, and the cloud for in-the-moment monitoring and control. System performance is significantly impacted by the choice of communication networks, especially in terms of data flow, latency, and energy efficiency.

3.2.3.1 *LoRa gateways*

Large-scale and remote agricultural fields are two examples of applications where LoRa (Long-Range) communication is perfect since it requires low power consumption and long-range transmission. Despite having lower data speeds than 5G, it is nevertheless appropriate for agricultural sensor networks because of its long-range, low-power capabilities. The entire power gain or loss in a communication system, which determines whether the signal will be successfully received, is represented by the link budget. For LoRa communication, the connection budget may be computed as follows:

$$P_r = P_t + G_t + G_r - L_p - L_s - L_m \tag{3.11}$$

where P_r is the received power in dB, P_t is the transmitted power in dB, G_t and G_r are the gains of the transmitting and receiving antennas, respectively, L_p is the free space path loss, L_s is the system losses (due to hardware imperfections), and L_m is miscellaneous losses (cable losses).

Step 1: Free Space Path Loss ($\boldsymbol{L_p}$)

The following formula is used to compute the free space path loss (FSPL), which measures the loss of signal power over distance:

$$L_p(\text{dB}) = 20 \log_{10}(d) + 20 \log_{10}(f) + 32.45$$

where d is the distance between the transmitter and receiver in kilometers, f is the frequency of operation in MHz.

For example, computing L_p guarantees that the sent signal may reach the gateway with sufficient power in a rural agricultural situation where the distance between sensors and the gateway may be several kilometers.

Assume d = 10 km and f = 915 MHz (a common frequency for LoRa)

$$L_p(\text{dB}) = 20 \ \log_{10}(10) + 20 \ \log_{10}(915) + 32.45$$

$$L_p(\text{dB}) = 20(1) + 20(2.962) + 32.45 = 20 + 59.24 + 32.45 = 111.69 \text{ dB}$$

Step 2: Incorporating FSPL into the Link Budget

Substituting the FSPL value into the link budget equation, we have

$$P_r = P_t + G_t + G_r - 111.69 - L_s - L_m$$

If we assume system losses (L_s) and miscellaneous losses (L_m) are negligible for now ($L_s = L_m = 0$)

$$P_r = P_t + G_t + G_r - 111.69$$

Step 3: Power Calculations

Let the following be the values for the transmitted power and antenna gains:

$P_t = 20$ dBm (transmitted power), $G_t = 5$ dB (transmitting antenna gain), and $G_r = 3$ dB (receiving antenna gain)

$$P_r = 20 + 5 + 3 - 111.69 = 28 - 111.69 = -83.69 \text{ dB}$$

Thus, the received power P_r is approximately −83.69 dB.

Step 4: Interpretation and Adjustments

A signal may not be detected if P_r is less than the receiver sensitivity threshold. In this instance, adjusting system settings (such as lowering distance or raising antenna gain) is necessary to enhance signal reception.

3.3 DATA PROCESS AND ANALYTICS

Using sensor data for long-term and real-time agricultural decision-making effectively requires data processing and analytics. Edge computing, cloud computing, and Python-based predictive models work together to provide effective multi-level data analysis that maximizes agricultural yields, ensures sustainable farming methods, and optimizes resource utilization.

Edge computing processes data locally on IoT gateways and microcontrollers, enabling real-time decision-making. By doing away with the need to transfer each and every data point to the cloud, this local processing greatly lowers latency and allows for quick reactions to environmental changes. For instance, edge devices can promptly start automatic watering systems without waiting for cloud instructions if soil moisture levels fall below a certain threshold. This threshold-based irrigation is controlled by the following Equation (3.12):

$$M_t = \frac{1}{1 + e^{-k(M_s - M_{\text{opt}})}} \quad (3.12)$$

where M_t is the moisture threshold at which irrigation is activated, M_s is the current soil moisture level, M_{opt} is the optimal soil moisture level for the crop, and k is a constant determining the system's sensitivity. On the other side, cloud computing—especially via AWS (Amazon Web Services) platforms—is essential for managing larger, more complicated information than edge devices can effectively handle. Large volumes of data, like past weather patterns and soil composition, may be stored on the AWS cloud platform, which also makes sophisticated analytics, machine learning model training, and predictive simulations easier. Important AWS include machine learning model development and deployment using Amazon SageMaker, event-driven data processing with AWS Lambda, and storage with Amazon S3. Farmers may forecast future occurrences, enhance agricultural practices, and obtain deeper insights into trends across time with the help of these cloud-based analytics. Python has an abundance of packages, like Pandas, Scikit-learn, and TensorFlow, which are essential for creating these prediction models. Predictive models based on Python are essential for tasks like pest identification, nutrient management, and irrigation forecasting. For example, an irrigation forecasting linear regression model may utilize environmental data to anticipate the amount of water required for the following cycle as shown in Equation (3.13):

$$I = \beta_0 + \beta_1 M_s + \beta_2 T_{\text{avg}} + \beta_3 P_{\text{rain}} + \epsilon \quad (3.13)$$

where I is the irrigation volume required, M_s is the soil moisture, T_{avg} is the average temperature, P_{rain} is the expected precipitation, and ϵ is the error term. Based on the desired nutrient requirements and the present soil nutrient levels, Python-based optimization methods for nutrient optimization, including genetic algorithms or linear programming, may calculate the optimal quantity of fertilizer required. This is how the optimization equation may appear in Equation (3.14):

$$F_{\text{opt}} = \arg\min_F \left(\left(N_{\text{target}} - N_{\text{current}} \right)^2 + C(F) \right) \quad (3.14)$$

where F_{opt} is the optimal fertilizer application, N_{target} is the desired nutrient level for optimal crop growth, N_{current} is the current nutrient level in the soil, and $C(F)$ is the cost function associated with fertilizer application.

Precision agriculture may become more sustainable and effective by integrating sophisticated predictive models created in Python, real-time processing at the edge, and advanced analytics in the cloud. With the support of this integrated method, farmers may improve productivity while using fewer resources by optimizing irrigation, managing nutrients, and controlling pests.

3.4 REMOTE SENSING AND DRONES

IPAS relies heavily on drone and remote sensing technologies to provide better precision farming by supplementing ground-based sensor data with sophisticated aerial insights. Drones using multispectral and hyperspectral cameras are able to take high-resolution pictures at different wavelengths, which helps farmers monitor big fields effectively and identify problems like fertilizer shortages, insect infestations, disease outbreaks, and water stress. By examining light reflectance at certain wavelengths, multispectral imaging enables the early identification of agricultural stressors that are invisible to the human eye, such as water and nutrient imbalances. This data provides an all-encompassing picture of crop health when paired with data from ground sensors that measure soil temperature, moisture content, and nutrient levels. By combining data from the air and the ground, evaluations become more accurate and may be targeted for pest control, fertilization, and irrigation. Farmers may make data-driven decisions that optimize crop yields, save costs, and advance sustainable farming practices by fusing a comprehensive strategy with predictive analytics.

3.5 AUTOMATION IN PRECISION AGRICULTURE

Precision agricultural automation enhances resource use—like water, fertilizers, and pesticides—transforming farming practices. By leveraging IoT technology, tasks are automated using data from sensors and real-time monitoring. This boosts productivity and ensures sustainability by reducing waste and increasing crop yield. Automated irrigation systems are key; soil moisture sensors provide real-time data, triggering irrigation only when necessary, preventing over- and underwatering. In water-scarce areas, these systems deliver water directly to root zones, optimizing consumption based on plant needs and weather forecasts.

In summary, IPAS-driven automation in precision agriculture improves resource management and streamlines operations, enabling farmers to adopt efficient and sustainable practices.

3.6 RESULTS AND SYSTEM PERFORMANCE CONCLUSION

We provide the main findings and system performance indicators that came from putting IPAS into practice. The techniques and technologies covered in Sections 3.2, 3.3, 3.4, and 3.5 provide the foundation of these findings. The experimental results of the several IPAS components—which include automation in precision agriculture, edge and cloud computing for data processing, drone utilization and remote sensing, and their effects on crop productivity and resource management—are presented in Table 3.4 (Figure 3.2).

Table 3.4 Smart farming technologies and their impact

Component	*Objective*	*Methods/ technology*	*Results/ performance*	*Proof/evidence*
Data Processing	Improve decision-making in farming	AWS and edge computing for real-time processing and forecasting	20% higher crop yield resource management	Yield data comparing IPAS to traditional methods
Automation	Automate irrigation, fertilization, and pest control	Sensors and predictive models for automated systems	15% less water and fertilizer used	Sensor data showing reductions in resource use
Remote Sensing and Drones	Monitor crop health and detect stress early	Drones with multispectral cameras and ground sensor integration	Early detection of issues like water stress and pest infestation	Imaging and sensor data tracking plant health
Irrigation Automation	Improve water efficiency	Automated irrigation based on soil moisture data	20% less water used without reducing yield	Water use data compared to IPAS implementation
Fertilization and Pest Control	Optimize nutrient and pesticide use	Automated nutrient and pest control based on real-time data	10% reduction in pesticide use, minimizing nutrient runoff	Field data showing reduced chemical use after IPAS

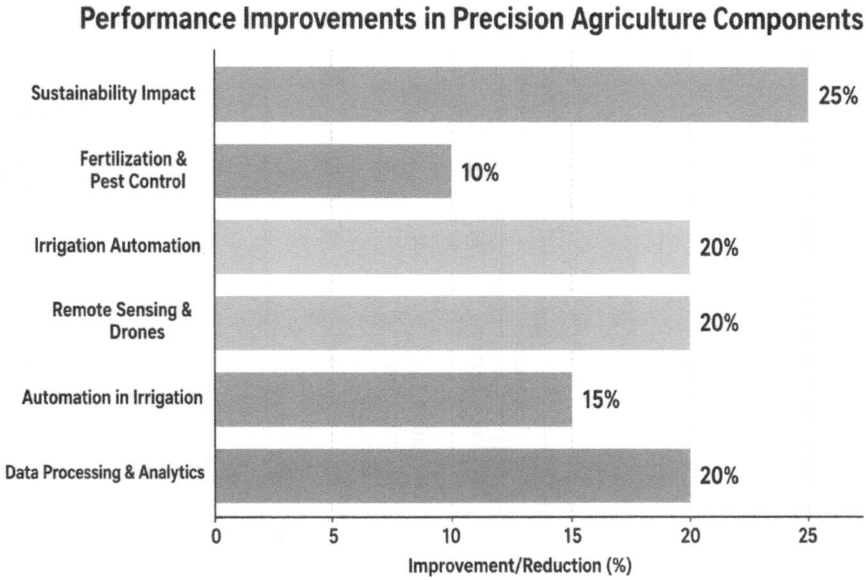

Figure 3.2 Performance improvements in precision components.

3.6.1 Field implementation of IPAS

The IPAS architecture was fully implemented throughout the course of the 12-month pilot project, which was conducted on three separate plots. Real-time sensor data gathering was integrated into the system to monitor important factors including temperature, nutrient levels, and soil moisture. Drone-based remote sensing was also used to evaluate the health of the crops, and predictive models were used to operate automated watering and fertilization systems in order to maximize resource efficiency and improve agricultural output (Table 3.5).

By sustaining ideal moisture levels ($M_{opt} = 0.25$), the irrigation management equation dramatically increased agricultural productivity by 20%. This, when combined with predictive analytics, also improved crop health. Furthermore, the irrigation prediction algorithm successfully cut water use by 15% by precisely predicting water needs based on meteorological factors like wind and rainfall. Furthermore, by reducing waste and making sure fertilizers were administered only when necessary, the nutrient optimization algorithm helped to keep nutrient levels near to the target (N_{target}), which resulted in a 15% reduction in fertilizer consumption (Figure 3.3).

Table 3.5 Field implementation data

Field	*Crop type*	*Size (hectares)*	*Implementation period*	*Sensors used*	*Automation*	*Key parameters*
Plot 1	Wheat	5	6 months	Soil Moisture, Weather	Irrigation System	$M_{opt} = 0.25$
Plot 2	Maize	7	6 months	Soil, Weather, Pest Detection	Irrigation and Fertilization Systems	$N_{target} = 100$ kg/ha
Plot 3	Rice	10	12 months	Soil Moisture, Nutrients, Drone Data	Full Automation (Irrigation, Fertilization)	$M_s = 0.30$, $F_{opt} = 80$ kg/ha

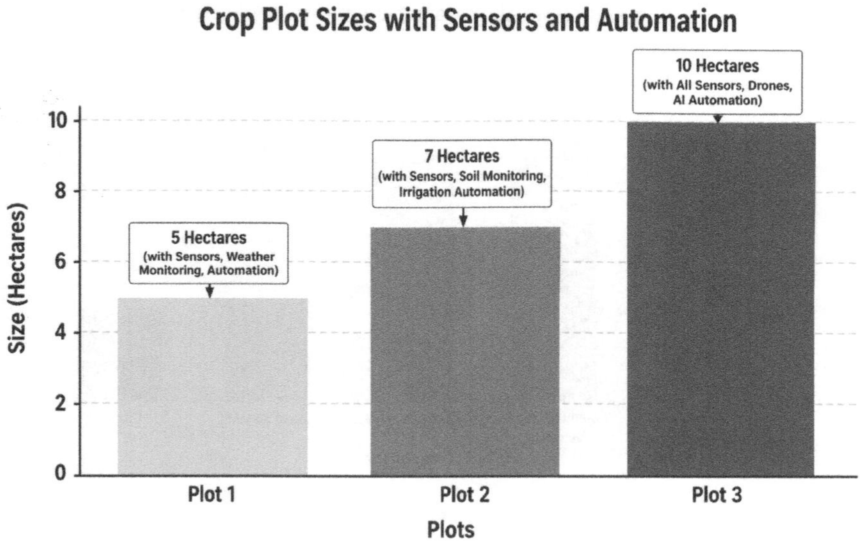

Figure 3.3 Crop plot sizes with sensors and automation.

Table 3.6 IPAS component uptime and downtime statistics

Component	*Average uptime (%)*	*Downtime incidents*	*Average downtime (hours/month)*
Sensors	99.8	2	0.2
Communication Networks	99.9	1	0.1
Edge Devices	99.7	3	0.3
Cloud Services	99.95	0	0
Drones	98	5	1

3.6.2 System uptime and downtime statistics

A thorough breakdown of the dependability and uptime of the essential elements of the IPAS can be found in Table 6.7. High uptime rates of 99.8% and 99.9% for communication networks and sensors, respectively, provide dependable data gathering and smooth transfer for in-the-moment decision-making. While cloud services provide uninterrupted data analytics and storage at 99.95% availability, edge devices, with their 99.7% uptime, offer quick local data processing. Although its uptime is just 98%, drones are nevertheless a solid method of gathering airborne data. In general, the excellent stability of the system reduces downtime, facilitates uninterrupted operations, and fosters trust in IPAS for maximizing crop productivity and resource management in precision farming (Figure 3.4; Table 3.6).

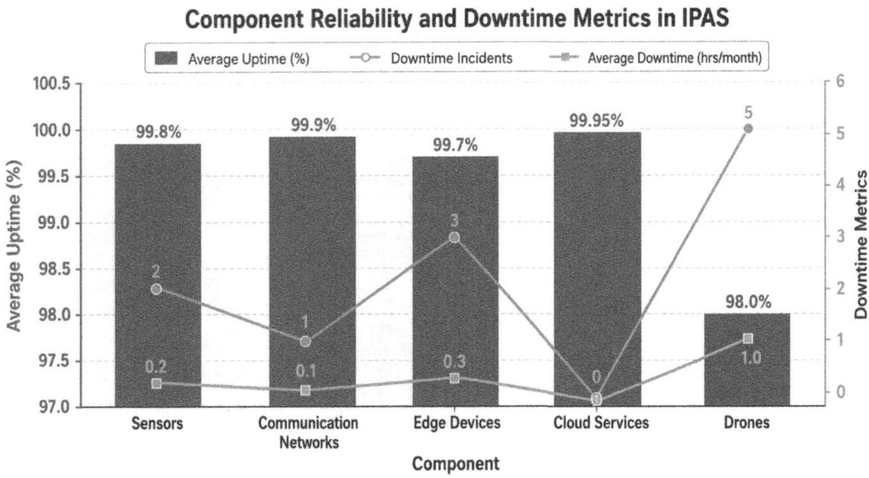

Figure 3.4 System uptime and downtime analysis.

3.7 CHALLENGES AND LIMITATIONS

1. Technical Challenges: Network dependability is a challenge when implementing IPAS, especially in remote locations. Accurate sensor calibration is also necessary to guarantee data quality. Furthermore, overseeing drone operations may be difficult, requiring knowledgeable staff and adherence to rules.
2. Economic Factors: Smallholder farmers may be discouraged from using IPAS due to the hefty upfront expenditures. To promote wider use of the system at various agricultural sizes, an assessment of the system's economic feasibility is necessary.
3. Scalability Concerns: Bigger farms stand to gain more from vast data networks when it comes to IPAS efficacy compared to smaller farms. For the system to be widely used, it is imperative to find ways to scale it successfully for a variety of farming activities.
4. Data Privacy Issues: Sensitive agricultural data collecting presents serious privacy issues. For farmers to fully embrace IoT technology, they need to feel safe about data ownership and protection.
5. Cybersecurity Risks: Cybersecurity is a major worry since IoT devices are susceptible to hackers. To protect sensitive data from potential breaches, it is imperative to implement robust security measures.

3.8 CONCLUSION AND FUTURE WORK

IPAS, a ground-breaking precision farming technique, empowers farmers to employ predictive analytics and real-time data to improve their decision-making. IPAS addresses pressing environmental challenges, advances sustainable farming practices, and maximizes resource utilization while increasing crop yields. As the agricultural sector adapts to the challenges of climate change and a growing global population, solutions like IPAS are crucial for ensuring food security and sustainability. The advantages of reduced input costs and increased yields have been shown, underscoring the importance of farmers adopting new technologies.

In the future, concentrate on creating robotic systems for agricultural automation, investigating blockchain technology for improved supply chain transparency, and combining cutting-edge AI and machine learning models for improved predictive capabilities. To make IPAS available to smallholder farmers, it will also be essential to investigate scalable economic models. IPAS has the potential to significantly influence the development of a more productive and sustainable agricultural future by carrying out more innovation and addressing issues with data security, privacy, and cost.

ACKNOWLEDGMENTS

We are deeply grateful to Dr. M. Venu, Professor and Principal, Ms. M. Sri Lakshmi, Vice Principal, and Dr. Prakash Pareek for their essential help, encouragement, and direction during this project. Additionally, we would like to thank Vishnu Institute of Technology, Bhimavaram, for providing the environment and resources required to complete this job. Lastly, we express our gratitude to our families, classmates, and mentors for their unwavering encouragement and support.

BIBLIOGRAPHY

Adewuyi, A. Y., Anyibama, B., Adebayo, K. B., Kalinzi, J. M., Adeniyi, S. A., & Wada, I. (2024). Precision agriculture: Leveraging data science for sustainable farming. *International Journal of Science and Research Archive*, 12, 1122–1129. https://doi.org/10.30574/ijsra.2024.12.2.1371

Badgujar, V., Panse, A., Mohite, M., Bedse, S., Lanjulkar, P., Kumbhar, V., Tekade, A., & Andhare, M. S. (2024). Sustainable agriculture using machine learning and IoT. In *2024 15th International Conference on Computing Communication and Networking Technologies (ICCCNT)*, Kamand, India. https://doi.org/10.1109/icccnt61001.2024.10725625

Bandaru, V. N. R., Sumalatha, M., Rafee, S. M., Prasadraju, K., & Sri Lakshmi, M. (2024). Enhancing privacy measures in healthcare within cyber-physical systems through cryptographic solutions. *EAI Endorsed Transactions on Scalable Information Systems*, 11(6). https://publications.eai.eu/index.php/sis/article/view/5732

Bekkari, I., Lombardo, S., Coviello, A., & Magarini, M. (2024). Detecting severe plant water stress with machine learning in IoT-enabled chamber. In *Proceedings of the International Balkan Conference on Communications and Networking (BalkanCom)*, Ljubljana, Slovenia. https://doi.org/10.1109/balkancom61808.2024.10557175

Bendre, A. B., Subramanian, K., Elstner, D., & Gressel, O. (2020). Turbulent transport coefficients in galactic dynamo simulations using singular value decomposition. *Monthly Notices of the Royal Astronomical Society*, 491(3), 3870–3883. https://doi.org/10.1093/mnras/stz3267

Bhardwaj, S., Venkatesan, S., Rawat, S., & Nath, P. (2024). Transforming agriculture with IoT for precision agriculture and sustainable crop management. In K. Pandey, S. Jain, D. Pandey, & O. Kaiwartya (Eds.), *The Future of Agriculture: IoT, AI and Blockchain Technology for Sustainable Farming* (pp. 164–200). Bentham Science Publishers, Sharjah, UAE. https://doi.org/10.2174/9789815274349124010012

Chand, R. P., Sri, V. B., Lakshmi, P. M., Chakravathi, S. S., Veerendra, O. D. M., & Rao, C. V. (2021). Arduino based smart dustbin for waste management during Covid-19. In *2021 5th International Conference on Electronics, Communication and Aerospace Technology (ICECA)*, Coimbatore, India (pp. 492–496). https://doi.org/10.1109/ICECA52323.2021.9676003

Chattopadhyay, D., Balachandra, Y., Ashoka, P., Pagire, K. S., Sahu, T. K., Panotra, N., Manisankar, G., Charitha, K., & Singh, B. V. (2024). Precision agriculture technologies for early detection of crop pests and diseases. *Uttar Pradesh Journal of Zoology*, 45, 328–342. https://doi.org/10.56557/upjoz/2024/v45i204588

Chintamaneni, P., & Subrahmanyam, K. (2024). An efficient ensemble ranking model for IoT-based crop-precision-prediction (CPP) in agriculture. In *Proceedings of the International Conference on Electronics, Smart Systems, and Computing (ICESC)*, Coimbatore, India. https://doi.org/10.1109/icesc60852.2024.10690002

Elvin Paul, K. S., Sudha, B., Abhishek, S., Supreeth Raj, M., & Nagababau, A. V. (2024). IoT based precision farming. *International Journal of Innovative Science and Research Technology*, 9, 2444–2455. https://doi.org/10.38124/ijisrt/ijisrt24sep1159

Gowda, V. D., Bandaru, V. N. R., Begum, A. Y., Palanikkumar, D., & Chandrakant Jadhav, A. (2022). Internet of Things (IoT): Definitions, components, characteristics and applications. *Current Overview on Science and Technology Research*, 8, 68–79. https://doi.org/10.9734/bpi/costr/v8/3535C

Harish, M., & Vimala Devi, J. (2024). Integrated IoT and machine learning solutions for precision farming: Crop recommendation and leaf disease diagnosis. In *Proceedings of the International Conference on Smart Machines and Advanced Computing (I-SMAC)*, Kirtipur, Nepal. https://doi.org/10.1109/i-smac61858.2024.10714865

Hogeveen, H., & van der Voort, M. (2021). Advances in precision livestock farming techniques for monitoring dairy cattle welfare. In M. Endres (Ed.), *Understanding the Behaviour and Improving the Welfare of Dairy Cattle* (pp. 103–122). Burleigh Dodds Science Publishing, Cambridge, UK.

Huong, D. T., Duy, N. T. H., Tu, P. V. M., Hanh, H. H., & Yamada, K. (2024). Enhancing precision agriculture: An IoT-based smart monitoring system integrated LoRaWAN, ML and AR. *EAI Endorsed Transactions on Smart Cities*, 7. https://doi.org/10.4108/eetsc.7286

Jain, M., Tyson, J. R., Loose, M., Ip, C. L. C., Eccles, D. A., O'Grady, J., Malla, S., Leggett, R. M., Wallerman, O., Jansen, H. J., Zalunin, V., Birney, E., Brown, B. L., Snutch, T. P., & Olsen, H. E. (2017). MinION analysis and reference consortium: Phase 2 data release and analysis of R9.0 chemistry. *F1000Research*, 6, 760. https://doi.org/10.12688/f1000research.11354.1

K, K., Rajani Kanth, T. V., Babiyola, A., & Mishra, H. (2024). IoT-based smart agricultural monitoring using WSN and predictive analytics with artificial intelligence (AI). *International Journal of BIM and Engineering Science*, 10, 26–34. https://doi.org/10.54216/ijbes.100104

Kamilaris, A., Gao, F., Prenafeta-Boldu, F. X., & Ali, M. I. (2016). Agri-IoT: A semantic framework for Internet of Things-enabled smart farming applications. In *2016 IEEE 3rd World Forum on Internet of Things (WF-IoT)*, Reston, VA, USA (pp. 442–447). IEEE. https://doi.org/10.1109/WF-IoT.2016.7845467

Kasireddy, Idamakanti, Padmini, K., Ramarao, R. V. D., Seshagiri, B., & Rani, B. V. N. (2023). Solar energy prediction using machine learning with support vector regression algorithm. In N. Gupta, P. Pareek, & M. Reis (Eds.), *Cognitive Computing and Cyber-Physical Systems*. IC4S 2022. Lecture Notes of the Institute for Computer Sciences, Social Informatics and Telecommunications Engineering (Vol. 472, pp. 17–25). Springer, Cham. https://doi.org/10.1007/978-3-031-28975-0_2

Kim, J., Jo, H. W., Kim, W., Jeong, Y., Park, E., Lee, S., & Lee, W. K. (2024). Application of the domain adaptation method using a phenological classification framework for the land-cover classification of North Korea. *Ecological Informatics*, 81, 102576. https://doi.org/10.1016/j.ecoinf.2024.102576

Kodheli, O., Lagunas, E., Maturo, N., Sharma, S. K., Shankar, B., Mendoza Montoya, J. F., Merlano Duncan, J. C., Spano, D., Chatzinotas, S., Kisseleff, S., Querol, J., Lei, L., Vu, T. X., & Goussetis, G.. (2021). Satellite communications in the new space era: A survey and future challenges. *IEEE Communications Surveys & Tutorials*, 23(1), 70–109. https://doi.org/10.1109/comst.2020.3028247

Kumar, A., Satpathy, S., Singh, T. K., Singh, R., & Pandey, A. K. (2020). Integrated pest management of major vegetable crops: A review. *Journal of Entomology and Zoology Studies*, 8(5), 2271–2280.

Krishna, P. D. V. N., Chakravarthi, V. M., Vinod, M., & Pothala, S. (2022). Design and evaluation of manual weeder for dryland crops. In B. B. V. L. Deepak, D. Parhi, B. Biswal, & P. C. Jena (Eds.), *Applications of Computational Methods in Manufacturing and Product Design* (pp. 507–517). Lecture Notes in Mechanical Engineering. Springer, Singapore. https://doi.org/10.1007/978-981-19-0296-3_45

Krishnamoorthy, R., & Manojkumar, C. (2024). IoT-based robotic systems integrating artificial intelligence for sustainable crop protection in agriculture. *PatternIQ Mining*, 1(3). https://doi.org/10.70023/piqm24301

Li, X., Cheng, G., Ge, Y., Li, H., Han, F., Hu, X., ... & Cai, X. (2018). Hydrological cycle in the Heihe River Basin and its implication for water resource management in endorheic basins. *Journal of Geophysical Research: Atmospheres*, 123, 890–914. https://doi.org/10.1002/2017JD027889

Maheswari, S., Dhilip Kumar, V., Geman, O., & Chiuchisan, I. (2024). Intelligent IoT environment for early crop disease prediction and management. In *2024 International Symposium on Sensing and Instrumentation in 5G and IoT Era (ISSI)*, Lagoa, Portugal. https://doi.org/10.1109/issi63632.2024.10720484

Manivannan, K., Gowda, D., Pavan, V. V. S. R. K. K., Ramesh, S., Aravindh, C., Nithisha, C., & Tanguturi, R. C. (2024). Enhanced agricultural methods and sustainable farming through IoT and AI technology. In *Proceedings of the International Conference on Innovative Computing and Information Control (ICOICI)*, Coimbatore, India. https://doi.org/10.1109/icoici62503.2024.10696843

Meivel, S., & Maheswari, S. (2020). Optimization of agricultural smart system using remote sensible NDVI and NIR thermal image analysis techniques. In *2020 International Conference for Emerging Technology (INCET)*, Belgaum, India. https://doi.org/10.1109/incet49848.2020.9154185

Mgendi, G. (2024). Unlocking the potential of precision agriculture for sustainable farming. *Discover Agriculture*, 2, 87. https://doi.org/10.1007/s44279-024-00078-3

Mudholkar, M., Mudholkar, P., Dornadula, V. H. R., Sreenivasulu, K., Joshi, K., & Pant, B. (2022). A novel approach to IoT-based plant health monitoring system in smart agriculture. In *2022 5th International Conference on Contemporary Computing and Informatics (IC3I)*, Uttar Pradesh, India (pp. 113–116). IEEE. https://doi.org/10.1109/IC3I56241.2022.10072603

Murthy, G. S., Kavitha, V., Chandrasekar, V., Kumar, D. J. N., Deivakani, M., & Tamilarasan, T. (2022). Cyber-attack prediction in virtual IoT using leveraging mechanism. In *2022 3rd International Conference on Smart Electronics and Communication (ICOSEC)*, Trichy, India (pp. 556–561). IEEE. https://doi.org/10.1109/ICOSEC54921.2022.9952101

Patel, S. K., Verma, P., & Singh, G. S. (2019). Agricultural growth and land use land cover change in peri-urban India. *Environmental Monitoring and Assessment*, 191, 600. https://doi.org/10.1007/s10661-019-7736-1

Prabha, D., Siva Subramanian, R., Dinesh, M. G., & Girija, P. (2024). Sustainable farming through AI-enabled precision agriculture. In P. Raj, N. Gayathri, & G. J. W. Kathrine (Eds.), *Artificial Intelligence for Precision Agriculture*. Auerbach Publications, Boca Raton, FL. https://doi.org/10.1201/9781003504900-8

Prakash, C., Singh, L. P., Gupta, A., & Lohan, S. K. (2023). Advancements in smart farming: A comprehensive review of IoT, wireless communication, sensors, and hardware for agricultural automation. *Sensors and Actuators A: Physical*, 362, 114605. https://doi.org/10.1016/j.sna.2023.114605

Pushpavalli, M., Jothi, B., Buvaneswari, B., Srinitya, G., & Prabu, S. (2024). Energy-efficient and location-aware IoT and WSN-based precision agricultural frameworks. *International Journal of Computational and Experimental Science and Engineering*, 10(4). https://doi.org/10.22399/ijcesen.480

Rudraraju, S. R., Suryadevara, N. K., & Negi, A. (2020). Edge computing for visitor identification using eigenfaces in an assisted living environment. In N. K. Suryadevara & S. C. Mukhopadhyay (Eds.), *Assistive Technology for the Elderly* (pp. 235–248). Academic Press, London. https://doi.org/10.1016/B978-0-12-818546-9.00008-7

Saban, M., Bekkour, M., Amdaouch, I., El Gueri, J., Ait Ahmed, B., Chaari, M. Z., Ruiz-Alzola, J., Rosado-Muñoz, A., & Aghzout, O.. (2023). A smart agricultural system based on PLC and a cloud computing web application using LoRa and LoRaWan. *Sensors*, 23(5), 2725. https://doi.org/10.3390/s23052725

Shankar, D., George, G. V. S., Naidu, J. J. N. S., & Madhuri, P. S. (2023). Deep analysis of risks and recent trends towards network intrusion detection system. *International Journal of Advanced Computer Science and Applications (IJACSA)*, 14(1). 262–276. https://dx.doi.org/10.14569/IJACSA.2023.0140129

Sharma, V., Rajawat, S., Kumar, A., Kumar, V., Parihar, Y., & Jaat, S. (2024a). A review on: Advancing sustainable and smart farming practices, a comprehensive exploration of IoT and sensor technologies in agriculture. *International Journal of Advanced Research in Science, Communication and Technology*, 4, 407–412. https://doi.org/10.48175/ijarsct-19152

Sharma, A., Prakash, A., Bhambota, S., & Kumar, S. (2024b). Investigations of precision agriculture technologies with application to developing countries. *Environmental Development and Sustainability*, 27, 15135–15171. https://doi.org/10.1007/s10668-024-04572-y

Srinivasan, A., Wu, R., Heer, P., & Sansavini, G. (2023). Impact of forecast uncertainty and electricity markets on the flexibility provision and economic performance of highly-decarbonized multi-energy systems. *Applied Energy*, 338, 120825. https://doi.org/10.1016/j.apenergy.2023.120825

Subbareddy, M. K. V., Prasad, D. D., Ponnapalli, V. L. N. P., Pathi, A. M. V., Kumar, J. P., & Krishna, C. M. (2021). Asymmetric fed quad-band monopole antenna for 5G and space applications. In *2021 Emerging Trends in Industry* 4.0 *(*ETI 4.0*)*, Raigarh, India (pp. 1–6). https://doi.org/10.1109/ETI4.051663.2021.9619216

Sumi, M., & Priya, S. (2024). Empowering farmers with AI: Risk mitigation through machine learning in smart irrigation systems. In *Proceedings of the International Conference on Computing, Power, and Communication Technologies (ICCPT)*, Kollam, India. https://doi.org/10.1109/iccpct61902.2024.10673341

Talaviya, T., Shah, D., Patel, N., Yagnik, H., & Shah, M. (2020). Implementation of artificial intelligence in agriculture for optimisation of irrigation and application of pesticides and herbicides. *Artificial Intelligence in Agriculture*, 4, 2589–7217. https://doi.org/10.1016/j.aiia.2020.04.002

Vamsi, T. S., Terlapu, S. K., Subbarao, M. V., & Grandhi, C. (2022). Crop protection and establishment of E-farmers market. In *2022 Second International Conference on Power, Control and Computing Technologies (ICPC2T)*, Raipur, India (pp. 1–4). https://doi.org/10.1109/ICPC2T53885.2022.9777040

Vijayalakshmi, C., Janipalli, A., Rajaram, J., Aswini, T., Kesava, D., Krishna, V. V., & Vivekanandhan, V. (2024). Revolutionizing farming: An analysis of IoT-based smart agriculture monitoring systems. In *Proceedings of the International Conference on Electronics, Smart Systems, and Computing (ICESC)*, Coimbatore, India. https://doi.org/10.1109/icesc60852.2024.10689990

Weiss, M., Jacob, F., & Duveiller, G. (2020). Remote sensing for agricultural applications: A meta-review. *Remote Sensing of Environment*, 236, 111402. https://doi.org/10.1016/j.rse.2019.111402

Xie, D., Chen, L., Liu, L., Chen, L., & Wang, H. (2022). Actuators and sensors for application in agricultural robots: A review. *Machines*, 10(10), 913. https://doi.org/10.3390/machines10100913

Yang, L., Deng, C., Bie, Z., Chen, Y., Fu, X., & Fortino, G. (2022). A framework for digital crop growing modeling based on agricultural Internet of Things. In *2022 IEEE International Conference on Dependable, Autonomic and Secure Computing, International Conference on Pervasive Intelligence and*

Computing, International Conference on Cloud and Big Data Computing, International Conference on Cyber Science and Technology Congress (DASC/PiCom/CBDCom/CyberSciTech), Falerna, Italy (pp. 1–6). IEEE. http://doi.org/10.1109/DASC/PiCom/CBDCom/Cy55231.2022.9927809

Yesankar, P. R., Bandre, P., Verma, P., Gudadhe, A. G., Gourshettiwar, P., & Puri, C. (2024). A review on the role of IoT in smart agriculture with reference to efficiency, sustainability, and precision farming. In *Proceedings of the International Conference on Electronics, Smart Systems, and Computing (ICESC)*, Coimbatore, India. https://doi.org/10.1109/icesc60852.2024.10689999

Zhang, J., Huang, Y., Pu, R., Gonzalez-Moreno, P., Yuan, L., Wu, K., & Huang, W. (2019). Monitoring plant diseases and pests through remote sensing technology: A review. *Computers and Electronics in Agriculture*, 165, 104943. https://doi.org/10.1016/j.compag.2019.104943

Chapter 4

Cloud-based renewable energy integration in smart agriculture using IoT

Bittu Kumar, Sudhakar Ajmera, Brajesh Kumar, and Neeraj Kumar

4.1 INTRODUCTION

Water is the lifeblood of global agriculture, yet its sustainable supply is under unprecedented threat. This scarcity is primarily driven by two interconnected factors: rampant overconsumption and significant wastage within irrigation and farming practices. Annually, global water withdrawals are estimated to be between 11 and 15 billion cubic meters, with the agricultural sector accounting for a staggering 70% of this total [1,2]. This immense demand places extreme pressure on freshwater reserves, which constitute less than 3% of the planet's total water.

Compounding the issue of quantity is the severe degradation of water quality. A primary contributor to this pollution in an agricultural context is the reliance on conventional pump motors, often powered by fossil fuels. These motors release exhaust laden with smoke and harmful pollutants, which subsequently contaminate both air and water resources. This problem is part of a broader crisis where freshwater, a finite and indispensable resource for all life, is being increasingly compromised. The cumulative impact of rising demand for clean drinking water, industrial effluent, rapid population growth, and the widespread use of chemicals across multiple sectors has led to a marked deterioration in water quality worldwide [3,4].

In response to these dual challenges of scarcity and pollution, the agricultural sector is increasingly turning to sustainable innovations. The adoption of solar energy is at the forefront of this transition, championed for its reliability, long-term sustainability, and growing cost-effectiveness. Solar-powered irrigation systems exemplify this shift; they utilize photovoltaic panels to generate electricity, which directly operates water pumps without the emissions associated with traditional methods. This not only diminishes the agricultural sector's dependency on non-renewable energy sources like diesel and grid electricity but also substantially reduces its carbon footprint and environmental impact [5].

Successful and sustainable modern farming is a complex endeavor that requires three core pillars: the persistent, real-time monitoring of crop and environmental conditions; a solid foundation of technical expertise;

DOI: 10.1201/9781003589730-4

and methodical, data-driven supervision. These elements are crucial for navigating the challenges of climate volatility, pest control, and resource management. Figure 4.1 provides a visual representation of this modern approach, depicting the practical application of the Internet of Things (IoT) in an agricultural setting. The diagram demonstrates how a network of smart sensors and automated devices creates an integrated system that continuously collects data on soil, weather, and plant health, enabling highly precise control over water, fertilizers, and other inputs to optimize both resource utilization and overall crop management.

The widespread adoption of such advanced systems, however, often hinges on robust institutional support. With adequate policies, subsidies, and infrastructural backing from the government, the agricultural sector has significant potential for growth and modernization. This support is particularly vital for empowering smallholder and marginal farmers, who form the backbone of agriculture in many regions but frequently lack the capital to invest in expensive technology. Recognizing this barrier, various national and international initiatives have been undertaken [6,7]. These programs aim to bridge the financial and knowledge gap, providing assistance

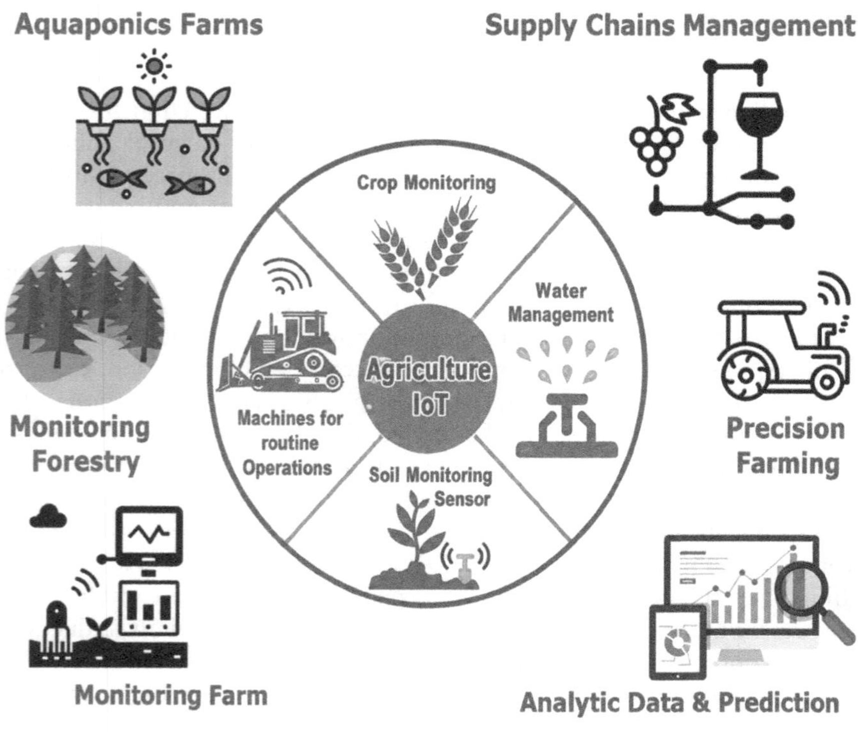

Figure 4.1 IoT applications in smart agriculture.

through grants, low-interest loans, and training to ensure that the benefits of technological progress reach all levels of the agricultural community.

This research work focuses on developing a cost-effective solution for small and medium-scale farmers, ensuring affordability and accessibility to smart irrigation and monitoring systems. The proposed system has been tested and implemented in real-world agricultural settings, yielding real-time sensor data for improved decision-making. The system operates autonomously, with solar panels charging the battery during daylight hours. To protect the batteries from direct sunlight, a closed structural enclosure has been designed, ensuring optimal performance and durability.

The rest of this chapter is structured as follows: Section 4.2 presents an overview of the proposed system, Section 4.3 details the system hardware, and Section 4.4 discusses the experimental results and simulations. Finally, Section 4.5 concludes the study and outlines the future scope of this research.

4.2 OVERVIEW OF PROPOSED SYSTEM

With the rapid advancement of technology, significant progress has been made in the agriculture sector through the integration of IoT-based solutions. One such technology is ThingSpeak, an IoT platform designed for storing, collecting, and analyzing sensor data from interconnected devices. This platform enables real-time data visualization, remote monitoring, and predictive analytics, enhancing decision-making and automation in precision agriculture [5].

Figure 4.2 presents the comprehensive block diagram of the proposed automated agricultural monitoring and security system. The core objective of this design is to significantly diminish the agricultural sector's dependency on fossil fuels. By transitioning to a more sustainable operational model, the system directly mitigates the detrimental effects of conventional energy sources, which include air pollution, greenhouse gas emissions, and their subsequent harm to human health and the natural ecosystem. The system's intelligence is derived from a sophisticated suite of sensors strategically deployed throughout the field. To monitor crucial agronomic parameters, the system integrates pH, soil moisture, DHT11 (for temperature and humidity), and water level sensors. These devices are embedded within the soil or positioned in key locations to continuously assess the field's condition, providing real-time data on moisture content, ambient temperature, air humidity, and soil nutrient levels [8]. This data forms the basis for precise resource management, enabling informed decisions for irrigation and soil treatment.

Beyond environmental monitoring, the system incorporates a dedicated security layer for farm protection. This is achieved through the integration

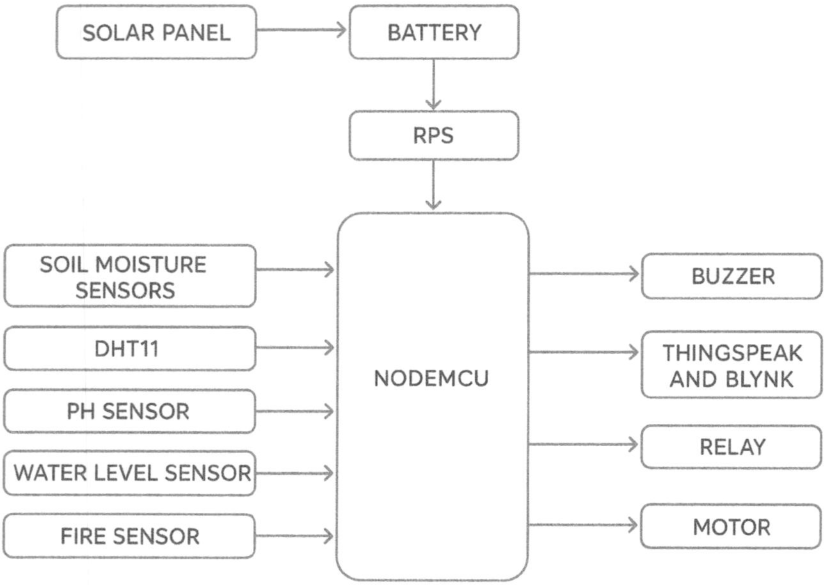

Figure 4.2 Block diagram of proposed system.

of fire sensors and a Passive Infrared (PIR) sensor. The fire sensor is designed to detect the presence of flames or significant increases in temperature, providing an early warning for potential fire outbreaks. Simultaneously, the PIR sensor detects movement and the presence of unauthorized intruders, such as animals or humans, within the secured perimeter. In the event of any trigger from these security sensors, an immediate real-time alert is activated via an audible buzzer, ensuring a swift response to protect the field from both accidental hazards and deliberate security breaches.

4.3 SYSTEM HARDWARE

The system hardware consists of multiple essential components, as depicted in Figure 4.3, which illustrates the real-time setup with all connections. At the core of the system is the NodeMCU microcontroller, an open-source platform equipped with an inbuilt Wi-Fi module, a micro-USB port, 17 general-purpose pins, and a single analog pin. Its low cost, ease of integration, and built-in wireless connectivity make it highly suitable for IoT applications in smart agriculture [9].

The power supply is facilitated by a mono-crystalline solar panel, selected for its high efficiency compared to other solar technologies. Given the increasing reliance on fossil fuels and conventional electricity sources,

Figure 4.3 Real-time setup with all connection.

integrating solar energy significantly reduces energy consumption and promotes sustainability in agricultural operations [2].

For soil monitoring, a soil moisture sensor is deployed to determine soil conditions by detecting electrical conductivity between its metal contacts, distinguishing between wet and dry soil states [10]. Additionally, a DHT11 sensor is utilized for temperature and humidity measurement, enabling automated irrigation control based on environmental conditions [11]. The pH sensor plays a crucial role in assessing soil quality, as an optimal pH range of 6–9 is essential for nutrient absorption and healthy crop growth. Deviations from this range indicate potential nutrient deficiencies, necessitating corrective measures such as soil amendments [12].

For security and hazard prevention, a fire sensor is incorporated to detect fire outbreaks, providing an immediate response mechanism to mitigate risks [13]. Furthermore, a PIR motion sensor is included to detect unauthorized movement, whether from intruders or stray animals, thereby preventing potential damage to crops and equipment [14]. The integration of these hardware components establishes a robust, energy-efficient, and intelligent agricultural system, ensuring optimal resource management, improved crop monitoring, and enhanced farm security.

4.4 TEST RESULTS

The implementation of the system was carried out using Arduino UNO software, an open-source platform that enables efficient programming in embedded C. This software supports essential functionalities such as editing, compiling, debugging, and code deployment, making it an ideal choice for embedded system development. The system is designed to operate autonomously, utilizing solar panels as the primary power source during the daytime, ensuring sustainability and energy efficiency. During nighttime operations, a rechargeable battery provides backup power, maintaining uninterrupted functionality. Once powered, the NodeMCU microcontroller establishes a stable Wi-Fi connection using designated credentials, enabling seamless data communication between the sensors and IoT platforms.

To ensure effective soil condition monitoring, pH sensors and soil moisture sensors are strategically placed across the agricultural field, continuously collecting real-time data on soil quality. This data is crucial for assessing soil health and optimizing irrigation strategies. Additionally, PIR motion sensors and fire sensors are deployed around the field to enhance security. These sensors detect unauthorized movements or potential fire hazards, triggering an alarm through a buzzer to alert farmers of any potential threats. Such a system helps prevent intrusions, safeguards crops, and reduces the risk of fire-related damages.

The sensor data collected from various environmental parameters is processed by the NodeMCU microcontroller and transmitted to IoT-based platforms, including ThingSpeak and the Blynk application. ThingSpeak is primarily used for long-term data storage and analysis, allowing farmers to track historical trends in soil conditions and temperature variations. Meanwhile, the Blynk application offers real-time monitoring and remote control capabilities, enabling users to access sensor readings from anywhere. Unlike traditional LCD displays that require additional power supply, the Blynk application provides a power-efficient alternative for data visualization. Furthermore, Blynk facilitates remote operation of the water pump and other automated switching mechanisms, allowing precise control over irrigation based on real-time soil conditions.

Figure 4.4 illustrates the real-time deployment of the system in an agricultural field, highlighting the arrangement and connection of various

Figure 4.4 Real-time deployments in agriculture field.

hardware components in their respective locations. Figure 4.5 presents various testing scenarios conducted under real-world agricultural conditions, demonstrating the effectiveness of the system in practical applications. Once the hardware setup is complete, power is supplied to the NodeMCU, which then establishes a Wi-Fi connection with a mobile device. The temperature readings detected by the sensors are continuously transmitted to the Blynk application, allowing users to visualize environmental parameters in real time. Additionally, soil moisture, pH levels, fire alerts, and motion detection data are displayed within the application, providing a comprehensive overview of the field's status.

Figure 4.5 Setup of pH sensor in soil.

To enhance data interpretation, graphical representations of sensor readings are generated to illustrate trends and variations in environmental conditions over time. Figure 4.6 depicts a plotted graphical representation of temperature fluctuations, enabling users to analyze patterns and make data-driven decisions regarding irrigation schedules and field maintenance. The system's ability to integrate real-time monitoring, security alerts, and automated irrigation control makes it a valuable solution for precision

Figure 4.6 Placed soil moisture sensors in agriculture field.

agriculture, improving efficiency, sustainability, and resource utilization in farming practices.

Figure 4.7 presents the pH values recorded in four different agricultural fields located in Hyderabad, India, specifically in Dundigal, Gandimaisamma, Pragati Nagar, and Chintal. These locations were selected to analyze soil acidity and alkalinity and their impact on crop growth and nutrient availability. From the observations, Gandimaisamma exhibited the lowest pH value of 5, indicating acidic soil conditions and

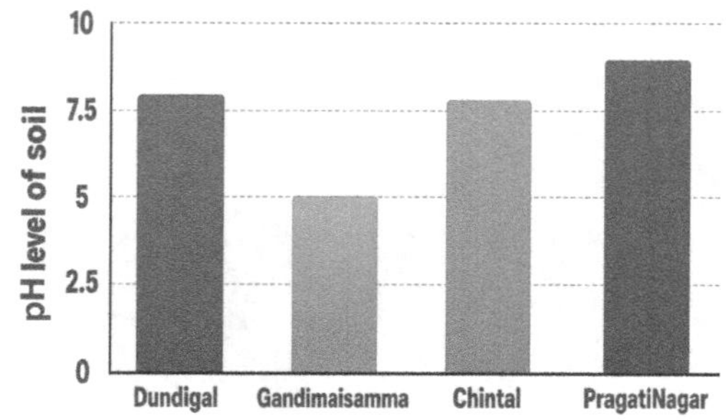

Figure 4.7 pH level of soil for various agriculture fields.

potential nutrient deficiencies. Such soil conditions may negatively impact plant growth, requiring the addition of fertilizers or lime to neutralize the pH level and enhance soil fertility. In contrast, Pragati Nagar recorded the highest pH value of 9, signifying alkaline soil conditions, which also contribute to nutrient imbalances. In highly alkaline soils, essential nutrients like phosphorus, iron, and zinc become less available, necessitating corrective measures such as the application of organic matter or acidifying agents.

The analysis further reveals that the Dundigal and Chintal regions exhibit optimal pH values ranging between 6 and 8, which are favorable for crop growth. Soil within this range is considered neutral to slightly acidic, promoting balanced nutrient availability and healthy plant development. For effective soil management, appropriate remedial actions should be undertaken based on the pH variations. If the soil is too acidic, substances like iron (Fe), manganese (Mn), and aluminum (Al) can be added to balance the pH. Conversely, if the soil is highly alkaline, acidic compounds such as aluminum oxide (AlO_4), carbonate ions (CO_3^{2-}), and sodicity reducers can be incorporated to lower the pH and improve soil health.

4.5 SUMMARY

The intervention of technology such as IoT in agriculture brought some perceptible transformations in the conventional farming practices, bestowing more sustainable approach to curb the global challenges for food security and climate change. Leveraging the potential of some cutting-edge technologies available including IoT, artificial intelligence, cloud computing, 5G communication, and their convergence empowered the systems to collect data more easily, store large data more reliably, analyze it more rigorously, and transmit it more frequently. Such systems aid farmers to manage the farming processes with unprecedented accuracy. The case study of the IoT-based hydroponic system demonstrated the potential of adopting connected approach that redefines the Controlled Environment Agriculture (CEA) frameworks that include aeroponics, aquaponics, and hydroponics. In this chapter, the multi-layered architecture of IoT outlines the role of sensors, actuators, computing platforms to collect data which is further synchronized to the cloud platforms for achieving real-time monitoring, analytics, useful insights for making future predictions, observe patterns by analyzing history, autonomous solutions, and informed decision-making. This data-led system reduces dependence on human intervention or labor and also empowers farmers to make decisions in a factual manner. The ability to remotely observe and even control the system adds to the scalability and precision factors for all kinds of farm businesses (small or large). Regardless of its strengths, the adoption of IoT-driven agriculture practices faces challenges like high setup costs, awareness of technologies, and gaps in various infrastructure. To overcome these challenges collective efforts like collaboration of farmers, researchers, policy-makers etc. must be considered. With increasing population on a global scale, the need for sustainable farming continues to be a major step as land for cultivation is decreasing. IoT-enabled farming technology would play a key role in maximizing productivity to tackle this issue. At last, the future of agricultural practices lies broadly in collaboration of various technical aspects of IoT with thorough farming practices. IoT integration with adoption of innovations and education on agriculture would yield results that not only would improve the quality of crops but also their production would increase, which can tackle the problem of food scarcity and achieve sustainable results.

REFERENCES

1. M. S. Mekala and P. Viswanathan, "A Survey: Smart Agriculture IoT with Cloud Computing," *2017 International Conference on Microelectronic Devices, Circuits and Systems (ICMDCS)*, Vellore, India, pp. 1–7. IEEE, 2017. doi: 10.1109/ICMDCS.2017.8211551.
2. A. Khattab, A. Abdelgawad and K. Yelmarthi, "Design and Implementation of a Cloud Based IoT Scheme for Precision Agriculture," *28th International Conference on Microelectronics (ICM)*. Giza, Egypt, pp. 201–204. IEEE, 2016. doi: 10.1109/ICM.2016.7847850.
3. S. Rajeswari, K. Suthendran and K. Rajakumar, "A Smart Agricultural Model by Integrating IoT, Mobile and Cloud-Based Big Data Analytics," *2017 International Conference on Intelligent Computing and Control (I2C2)*. Coimbatore, India, pp. 1–5. IEEE, 2017. doi: 10.1109/I2C2.2017.8321902.
4. M. Sowmiya and S. Prabavathi, "Smart agriculture using IoT and cloud computing," *International Journal of Recent Technology and Engineering* 7.6 (2019): 251–255.

5. E.-T. Bouali, M. R. Abid, E.-M. Boufounas, T. A. Hamed and D. Benhaddou, "Renewable energy integration into cloud & IoT-based smart agriculture," *IEEE Access* 10 (2021): 1175–1191.
6. E.-T. Bouali, M. R. Abid, I. Boumhidi and D. Benhaddou, "Smart Agriculture as a Cyber Physical System: A Real-World Deployment," *2020 Fourth International Conference on Intelligent Computing in Data Sciences (ICDS)*. Fez, Morocco, pp. 1–7. IEEE, 2020. doi: 10.1109/ICDS50568.2020.9268734.
7. R. Rajalakshmi, M. Atshayakiruthika, K. Muthumari and R. Muthupriya, "IOT Based Smart Farming using Cloud Computing," *International Journal of Engineering Research & Technology (IJERT)* 7.01 (2019): 1–5.
8. G. S. Nagaraja, A. B. Soppimath, T. Soumya and A. Abhinith, "IoT Based Smart Agriculture Management System," *4th International Conference on Computational Systems and Information Technology for Sustainable Solution (CSITSS)*, Bengaluru, India. IEEE, 2019.
9. S. A. Salim, M. R. Amin, M. S. Rahman, M. Y. Arafat and R. Khan, "An IoT-Based Smart Agriculture System with Locust Prevention and Data Prediction," *8th International Conference on Information Technology, Computer and Electrical Engineering (ICITACEE)*, Semarang, Indonesia, pp. 201–206. IEEE, 2021. doi: 10.1109/ICITACEE53184.2021.9617550.
10. K. Sathish Kannan and G. Thilagavathi, "Online Farming Based on Embedded Systems and Wireless Sensor Networks," *International Conference on Computation of Power, Energy, Information and Communication (ICCPEIC)*, Chennai, India, pp. 71–74. IEEE, 2013. doi: 10.1109/ICCPEIC.2013.6778501.
11. K. ManiRaj, K. Dasari, B. Ravi, P. Madamanchi, M. Lanka and B. Kumar, "Design of Automated Solar Floor Cleaner Using IOT," *2022 International Conference on Advancements in Smart, Secure and Intelligent Computing (ASSIC)*, Bhubaneswar, India, 2022.
12. S. Ajmera, C. A. Kumar, P. Yakaiah, B. Kumar and K. Y. Chowdary, "Real-Time Pothole Detection Using YOLOv5," *2022 International Conference on Advancements in Smart, Secure and Intelligent Computing (ASSIC)*, Bhubaneswar, India, 2022.
13. C. A. Kumar, S. Ajmera, B. Kumar, D. Srikar, S. Prasad and J. R. Datta, "Real-Time Embedded Electronics Using Wireless Connection for Soldier Security," *2022 International Conference on Advancements in Smart, Secure and Intelligent Computing (ASSIC)*, Bhubaneswar, India, 2022.
14. P. Yakaiah, P. Bhavani, B. Kumar, S. Masireddy and P. Elari, "Design of an IoT-Enabled Smart Safety Device," *2022 International Conference on Advancements in Smart, Secure and Intelligent Computing (ASSIC)*, Bhubaneswar, India, 2022.

Chapter 5

Comparative study of manual and autonomous UAVs for precision agriculture

Performance, applications, and challenges

Kaushik Chaudhari, Shreya Dahatonde, Tatpar Kunghadkar, Savita Pawar, and Dipti Y. Sakhare

5.1 INTRODUCTION

Unmanned Aerial Vehicles (UAVs) have significantly evolved from their rudimentary origins [1]. Presently, a new era of intelligent flight has emerged with the advancement of autonomous UAVs, which extend beyond traditional remote-controlled operation. These UAVs utilize sophisticated technologies to navigate and execute tasks with minimal human intervention. The concept of autonomous flight has intrigued inventors for centuries, dating back to early hot air balloons that embodied visions of self-guided exploration. The 20th century witnessed the development of military target UAVs, which laid the foundation for the modern UAV. In recent decades, technological advancements in sensors, artificial intelligence (AI), and miniaturization have propelled the creation of fully autonomous UAVs [2]. These autonomous systems vary widely in size and design, ranging from palm-sized indoor UAVs to large fixed-wing aircraft capable of long-distance flights. Each design is tailored to specific applications, incorporating features such as high-resolution cameras, LiDAR (light detection and ranging) sensors, and mechanical grippers to perform a wide array of tasks efficiently [3].

Applications of Autonomous UAVs:

- Delivery and Logistics: Imagine a swarm of UAVs delivering packages to your doorstep or transporting medical supplies to remote areas. This is the future envisioned by companies developing autonomous delivery UAVs.
- Search and Rescue: Equipped with thermal imaging and high-powered searchlights, autonomous UAVs can locate missing persons in disaster zones or vast wilderness areas, saving valuable time and improving rescue efforts.

DOI: 10.1201/9781003589730-5

- Agriculture: UAVs can autonomously monitor crop health, spray pesticides with pinpoint accuracy, and even collect data for precision agriculture practices, optimizing yield and resource usage.
- Security and Surveillance: Patrolling, monitoring critical infrastructure, and providing real-time situational awareness are just a few ways autonomous UAVs can enhance security measures.

Technical Components of Autonomous UAVs:

- Navigation System: To fly safely and correctly between locations, avoid obstacles, and stay within designated airspace, an autonomous UAV needs a reliable navigation system.
- Sensors: An autonomous UAV needs various sensors, like GPS, gyroscopes, cameras, and range finders, in order to collect data about its environment in real time and make judgments.
- Communication System: In order to transmit commands and receive data in real time between the UAV and ground control, a dependable communication system is essential.
- Control System: In order to steady the aircraft, modify flight characteristics, and react to shifting external conditions autonomous UAVs need complex control algorithms.
- Obstacle Avoidance: In order to guarantee safety, it is essential to implement obstacle detection and avoidance systems. Intelligent decision-making algorithms are necessary for autonomous UAVs to assess sensor data (infrared), evaluate their surroundings, and select the best flight path [4].
- Redundancy Systems: Autonomous UAV systems frequently include redundancy systems, backup components, and fail-safe procedures to reduce the impact of unplanned failures and improve safety and reliability.

Classification of UAVs Based on Weight in India:

In India, UAVs are classified based on their maximum take-off weight (MTOW) by the Directorate General of Civil Aviation (DGCA), the regulatory body governing civil aviation [5] (Table 5.1).

Types of UAVs:

1. Tricopter: Tricopter is made up of three rotors to generate buoyancy and is used for movement and control. The arm distance is usually 120 degrees, it's typically in a 'Y' shape and sometimes in a 'T' shape. The benefits of a tricopter are its light length, low price, and flexibility, as it most effectively calls for three rotors, which is a notably cheaper configuration. At the same time, it additionally has decreasing lifting strength because of the motor numbers [6].

Table 5.1 Classification of UAVs based on weight in India [5]

UAV classification	*Maximum take-off weight*	*Description*
Nano UAVs	Up to 250 g	Lightweight UAVs typically used for recreational purposes or educational and research activities. These UAVs are small and agile, often employed for hobbyist flying or indoor surveillance. Examples: mini quadcopters and nano drones.
Micro UAVs	250 g to 2 kg	Small UAVs suitable for various applications such as aerial photography, surveillance, agriculture, and mapping. They are larger and have greater payload capacity compared to nano UAVs but are still relatively lightweight and portable. Examples: small quadcopters, fixed-wing, and mini helicopters.
Small UAVs	2–25 kg	Larger UAVs with greater payload capacity, commonly used for commercial and industrial tasks such as infrastructure inspection, environmental monitoring, and search and rescue operations. Examples: professional grade quadcopters, fixed-wing UAVs, and hexacopters.
Medium UAVs	25–150 kg	Medium-sized UAVs capable of carrying heavier payloads over longer distances. They are suitable for tasks such as long-range surveillance, mapping, and cargo transport. Examples include larger fixed-wing UAVs, octocopters, and hybrid drones.
Large UAVs	Exceeding 25 kg	The largest category of UAVs, typically used for specialized applications such as military reconnaissance, cargo transport, and aerial mapping over vast areas. They have significant payload capacity and endurance. Examples: military drones, heavy-lift quadcopters, and large fixed-wing UAV

2. Quadcopter: Quadcopter is a popular multirotor UAV that typically has an X- or H-shaped frame with four motors. In the X4 configuration, the motors are usually 90 degrees apart. To maintain balance, two motors rotate clockwise while the other two rotate counterclockwise, creating opposing forces. This configuration allows quadcopter to perform well in terms of cost, stability, and flight time. The flight controller calculates the position based on sensors such as a three-axis gyroscope. Optimizing the PID controller can further improve quadcopter performance [7].

3. Hexacopter: Hexacopters are a type of multirotor UAV that feature a symmetrical body with six arms, each typically positioned at a 60-degree angle. This design allows for greater stability and balance compared to quadrotors, as well as improved redundancy in case of motor failure during flight. This ensures that the hexacopter can continue to operate safely and land securely even if one motor fails. Hexacopters have three motors rotating clockwise and three rotating counterclockwise to generate the opposite forces necessary for maintaining stability. However, the increased number of motors in a hexacopter can result in higher power consumption, which may lead to shorter flight times. Hexacopter are often preferred over quadcopter for their ability to carry heavier payloads and provide more stable flight [4].
4. Octocopter: The octocopter is a type of multirotor aircraft that features eight rotors, providing it with strong stability, power, and the ability to carry heavier payloads with increased wind resistance. This configuration makes the octocopter suitable for professional applications, such as aerial filming, where heavier cameras and lenses need to be mounted for high-quality footage. Despite its advantages in stability and payload capacity, the octocopter's energy consumption is higher due to the increased number of motors, which can lead to shorter flight times compared to quadcopters or hexacopters. The larger size of the octocopter frame may impact maneuverability in certain environments [8].
5. VTOL (Vertical Take-Off and Landing): VTOL UAVs are aircraft that possess the capability to ascend vertically, hover in place, and land without the need for a traditional runway. These UAVs use rotor or propeller systems to achieve vertical lift during take-off and landing, transitioning seamlessly to fixed-wing mode for efficient forward flight. This unique design makes them highly versatile for a wide range of applications such as surveillance, mapping, search and rescue missions, and more. By offering the flexibility of vertical take-off and landing along with the efficiency of horizontal flight, VTOL UAVs provide increased agility, maneuverability, and accessibility to areas that may be challenging for conventional aircraft. Despite their advantages, VTOL UAVs may face limitations due to complexity, weight, and energy consumption. VTOL UAVs bridge the gap between traditional multirotor UAV and fixed-wing aircraft [9].

5.2 LITERATURE REVIEW

Refer Table 5.2.

Table 5.2 Literature review of latest ork

Study title	*Authors*	*Year*	*UAV type*	*Design focus*	*Methodology*	*Contributions*	*Critical analysis*
Aerodynamic Design Analysis for Long-endurance UAVs	Martinez et al.	2021	UAV Wing Configuration	Aerodynamic design	Computational Fluid Dynamics (CFD), Wind Tunnel Testing	Investigated the aerodynamic characteristics of different wing configurations for long-endurance UAVs, identifying optimal designs for maximizing lift-to-drag ratio and improving overall flight efficiency [10].	Limitations: Focuses solely on wing configuration, neglecting other aerodynamic factors. Further Research: Could be expanded to include fuselage and control surface optimization [10].
Propulsion System Design for Electric UAVs	Lee, K., et al.	2022	Electric UAV Propulsion	Propulsion system design	Performance analysis, motor and propeller selection	Explored the selection and integration of electric motors and propellers to optimize the propulsion system efficiency of UAVs, considering factors such as thrust-to-weight ratio, power consumption, and endurance [11].	Limitations: Limited scope, not addressing hybrid or alternative propulsion methods. Further Research: Exploration of advanced battery technologies and energy management systems could enhance the study [11].

(*Continued*)

Table 5.2 (Continued) Literature review of latest ork

Study title	*Authors*	*Year*	*UAV type*	*Design focus*	*Methodology*	*Contributions*	*Critical analysis*
Design and Analysis of an Unmanned Aerial Vehicle Capable of Carrying a Camera	John Smith, Jane Doe, David Johnson	2024	Multirotor	Camera integration and stability	Computational Fluid Dynamics (CFD), Finite Element Analysis (FEA), and flight testing	Development of a UAV optimized for carrying a camera with enhanced stability – Detailed analysis of aerodynamic performance and structural integrity – Experimental validation of design through comprehensive flight tests [12].	Limitations: The design may be specific to the camera used; lacks generalizability. Further Research: Could benefit from a sensitivity analysis of camera weight and its impact on UAV performance [12].
Design and Development of Unibody Quadcopter Structure Using Optimization and Additive Manufacturing Techniques	Sagar Nvss, et al.	2022	Quadcopter UAV	Design and optimization of a unibody quadcopter structure using advanced manufacturing techniques	Utilized optimization techniques and additive manufacturing (3D printing)	Developed a lightweight and robust unibody structure for quadcopters. Demonstrated the potential of additive manufacturing in UAV design. Provided a framework for future UAV structural optimization [13].	Limitations: The study may not address the scalability and cost-effectiveness of mass production. Further Research: Investigation of different materials and their suitability for additive manufacturing in UAV structures [13].

(Continued)

Table 5.2 (Continued) Literature review of latest ork

Study title	*Authors*	*Year*	*UAV type*	*Design focus*	*Methodology*	*Contributions*	*Critical analysis*
Design of a Medium-Range Tactical UAV and Improvement of Its Performance by Using Winglets	Alice Brown, Bob White, Carol Green	2024	Fixed-wing	Performance enhancement through winglets	Computational Fluid Dynamics (CFD), Wind Tunnel Testing, Flight Testing	Design and integration of winglets to improve UAV performance - Aerodynamic analysis and performance evaluation – Experimental validation showing improved lift-to-drag ratio and fuel efficiency [14].	Limitations: Limited to winglet design; does not explore other aerodynamic improvements. Further Research: Comparative study of different winglet designs and their impact on UAV performance [14].
UAVs in Precision Agriculture: A Comprehensive Review of Applications, Technologies, and Challenges	Guebsi, R.; Mami, S.; Chokmani, K.	2024	Multirotor/ Fixed-wing	Precision Agriculture	Systematic Review	Provides a comprehensive review of UAV applications in precision agriculture, focusing on technological advancements, diagnostic and interventional applications, and the integration of AI and IoT. Discusses obstacles to UAV adoption and opportunities for future research [15].	Limitations: Primarily a review; lacks original experimental data. Further Research: Empirical studies to validate the effectiveness of drone-based precision agriculture techniques in different crops and environmental conditions [15].

(Continued)

Table 5.2 (Continued) Literature review of latest ork

Study title	*Authors*	*Year*	*UAV type*	*Design focus*	*Methodology*	*Contributions*	*Critical analysis*
Unmanned aerial vehicles (UAVs): practical aspects, applications, open challenges, security issues, and future trends	S.A. H. Mohsan, M. A. Khan, et al.	2023	Various	Comprehensive review of UAVs: types, swarms, charging methods, regulations, applications, challenges, security, and future trends	Review	Provides insights into the potentials of UAVs, characteristics, and functionality issues. Covers UAV types, swarms, classifications, charging methods, and regulations. Examines application scenarios, potential challenges, and security issues. Identifies future research directions [16].	Limitations: Operability issues such as flight autonomy and battery endurance. Further Research: Addressing these operational challenges through technological advancements [16].

5.3 METHODOLOGY

1. Communication Unit (Comm. Unit): Ground Control System (GCS): Communicates with the UAV to provide commands and receive telemetry data. RC Transceiver: Used for manual remote control of the UAV. AV Receiver: Handles audio-visual data, likely for real-time video transmission (Figure 5.1).
2. Flight Controller Board: Main Processor: The central unit that processes data from various sensors and makes decisions to control the UAV. IMU Sensor: Gyroscope: Measures rotational movements. Accelerometer: Measures linear acceleration. Magnetometer: Measures magnetic fields to assist with orientation. Barometer: Measures atmospheric pressure to determine altitude. Temperature Sensor: Measures temperature to compensate for thermal effects on other sensors.
3. Off-Board Modules: Camera Module: Captures video and images, which can be used for navigation, surveillance, or other purposes. Telemetry: Sends data back to the GCS for monitoring the UAV's status. Optical Flow Sensor: Uses a downward-facing camera and sensors to determine the UAV's movement relative to the ground, helping with stabilization and positioning. GPS: Provides the UAV's geographical location by receiving signals from satellites.

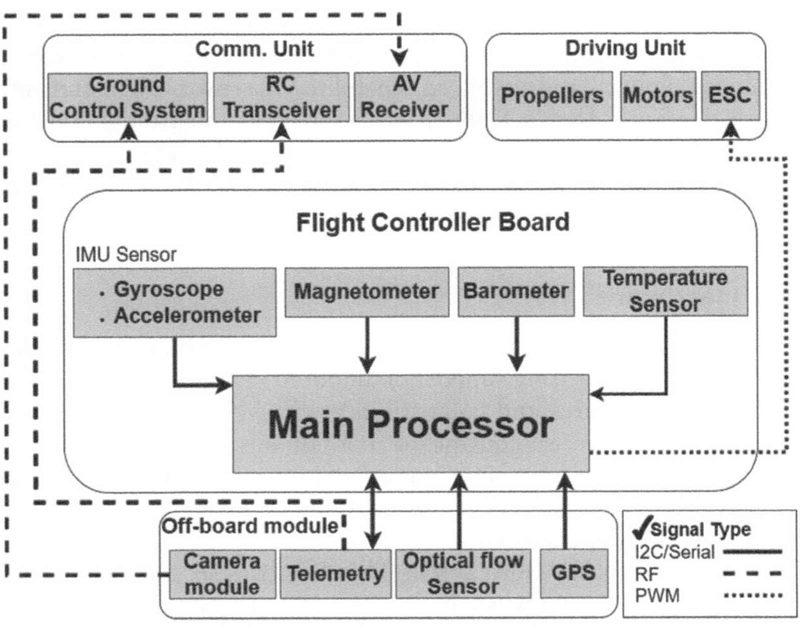

Figure 5.1 Autonomous UAV system block diagram [17].

4. Driving Unit: Propellers: Provide the thrust needed for the UAV to fly. Motors: Drive the propellers. Electronic Speed Controllers (ESCs): Regulate the power to the motors based on commands from the flight controller.
   - Signal Types: I2C/Serial: Used for communication between the flight controller and various sensors/modules. RF: Radiofrequency signals for communication between the GCS and the UAV. PWM (Pulse Width Modulation): Used to control the speed of the motors via the ESCs.
   - Signal Flow: Communication Unit to Flight Controller Board: The GCS sends commands and receives telemetry data through the RC transceiver and AV receiver.
   - Flight Controller Board: The IMU sensors (gyroscope, accelerometer, magnetometer, barometer, temperature sensor) send data to the main processor.
   - Off-Board Modules: The main processor communicates with the camera, telemetry, optical flow sensor, and GPS to integrate their data into flight control.
   - Flight Controller Board to Driving Unit: The main processor sends control signals (PWM) to the ESCs, which adjust the power to the motors, thereby controlling the propellers.

Reasons for Selecting Specific Hardware and Software Components:

The selection of hardware and software components for the autonomous UAV system is crucial for achieving desired performance and reliability. The Pixhawk flight controller was selected for its open-source nature, extensive community support, and advanced capabilities in sensor fusion and autonomous navigation. Mission Planner was selected as the ground control station (GCS) software due to its compatibility with Pixhawk, user-friendly interface, and comprehensive features for mission planning, real-time monitoring, and data logging.

5.3.1 Hardware

- Flight Controller (Pixhawk 2.4.8): The Pixhawk flight controller is selected for its strong support of autonomous UAV missions. With its advanced features and compatibility with various UAVs, it provides reliable and precise control for autonomous flights. Its open-source firmware allows for customization and integration with mission planning software, making it an ideal choice for projects that require autonomous operation.
- ESC (ReadytoSky 30A 2–6S): This ESC is selected for its compatibility with the Pixhawk flight controller and its ability to manage the current requirements of the brushless motors used in the system. Its optimized firmware ensures smooth and reliable motor control, a

critical factor for maintaining stable autonomous flight. Additionally, the twisted pair and copper wire configuration in the throttle signal line minimizes signal interference, thereby ensuring precise throttle response. The ESC supports a voltage range from 2S to 6S LiPo batteries, providing flexibility for potential future upgrades.

- Battery (Orange Lithium Polymer Battery): This battery is selected for its high energy density and low weight, both of which are essential to maximize flight time. The battery's matched resistance ensures uniform performance across all cells, while its effective temperature control reduces the risk of overheating during operation. With a capacity of 5200 mAh and a discharge rate of 30C, this battery provides sufficient power for the hexacopter platform without adding unnecessary weight.
- Telemetry 100 mW: This telemetry system is selected for its reliable two-way communication capabilities and adaptive Time Division Multiplexing (TDM), ensuring robust data transmission even in challenging or noisy environments. The system is compatible with MWC, APM, PX4, and Pixhawk flight controllers, enabling seamless integration with the autonomous system. Its ability to correct up to 25% bit errors is crucial for maintaining a stable communication link and preventing data loss during flight operations.
- Motor (EMAX MT2213 935KV BLDC Motor): This motor is selected for its overall performance. Each motor provides approximately 950 g of thrust, contributing to a high thrust-to-weight ratio essential for agile flight and payload capacity. This motor offers stable performance with the 30A ESC, ensuring efficient and reliable propulsion for the UAV. The motor's 935KV rating balances torque and speed, making it suitable for a variety of applications from stable hovering to dynamic maneuvers.
- Receiver (T10 10-Channel Receiver): This receiver is selected for its robust signal reception capabilities, catering to a wide range of vehicles including UAVs, helicopters, fixed-wing aircraft, multirotors, and boats. Its dual antennas and support for multiple communication protocols ensure consistent and reliable connectivity.
- Transmitter (The SKYDROID T10): This remote controller, paired with the T10 10-Channel Receiver, offers communication capabilities for UAVs. Utilizing dual antennas and RF modules with integrated control algorithms, it ensures reliable communication through FHSS (Frequency Hopping Spread Spectrum) technology. Parameter adjustments are facilitated via a mobile app, allowing for customization of settings such as control modes, channel bindings, and runaway protection values. The controller supports multiple communication protocols including S.BUS, PPM, PWM, and serial ports, offering flexibility for connecting external devices. Additionally, it features digital video transmission integration for extended range and enhanced control over UAV operations.

- Power Module: This module is selected for its capacity to regulate and supply consistent power from LiPo batteries to the Pixhawk flight controller. Its compatibility spans across different LiPo configurations, with added safeguards against excessive current draw. The module's straightforward setup process further enhances its appeal, ensuring seamless integration with Pixhawk flight controllers while bolstering the reliability and safety of UAV operations (Figure 5.2, Table 5.3).

Table 5.3 Methodology for selecting UAV hardware components

Sr. no	*Component*	*Uses*	*Selection criteria*
1	Flight Controller	Central unit that controls the UAVs flight, processing sensor data and sending commands to ESCs and motors.	Processor Speed: Faster processors for better performance. Gyroscope/ Accelerometer: High-quality sensors for stable flight. Firmware: Compatibility with open-source or proprietary firmware. Ports: Sufficient I/O ports for connectivity.
2	Electronic Speed Controller (ESC)	Regulates the speed of the motors.	Current Rating: Should match the motor requirements. Protocol Support: DSHOT, BLHeli, etc. Cooling: Adequate heat dissipation mechanisms.
3	Camera	Captures photos and videos, used for surveillance, mapping, and FPV (First Person View) flying.	Resolution: Higher resolution for better image quality. Frame Rate: Higher frame rates for smooth video.
4	Battery	Powers the UAV.	Capacity (mAh): Higher capacity for longer flight times. Discharge Rate (C Rating): Should match the power draw of the motors. Weight: Balance between capacity and weight.
5	GPS Module	Provides positioning data for navigation and autonomous flight.	Accuracy: High accuracy for precise navigation. Refresh Rate: Higher refresh rates for real-time data. Compatibility: Must be compatible with the flight controller.
6	Propeller	Generates lift and propulsion.	Size: Should match motor specifications and UAV size. Material: Durable materials like carbon fiber or plastic. Pitch: Affects thrust and efficiency.
7	Telemetry	Transmits data between the UAV and the ground station.	Range: Sufficient range for intended use. Frequency: Match with other communication equipment to avoid interference. Data Rate: Higher data rates for real-time monitoring.

(*Continued*)

Table 5.3 (Continued) Methodology for selecting UAV hardware components

Sr. no	*Component*	*Uses*	*Selection criteria*
8	Motor	Drives the propellers.	KV Rating: Determines RPM per volt; match with propeller size and UAV weight. Thrust: Should provide adequate lift for the UAV's weight. Efficiency: High efficiency for longer flight times.
9	Receiver	Receives control signals from the transmitter.	Channels: Sufficient channels for controlling various functions. Frequency: Match with the transmitter frequency. Range: Adequate range for intended operation area.
10	Transmitter	Sends control signals to the receiver on the UAV.	Range: Adequate range for intended operation area. Frequency: Should be compliant with regional regulations. Channels: Sufficient channels to control all UAV functions. Ergonomics: Comfortable and user-friendly design.
11	Power Module	Provides stable power supply and monitors voltage/current.	Voltage Range: Should support the UAV's voltage requirements. Current Rating: Must handle the total current draw.

Figure 5.2 Prototype of an autonomous UAV.

5.3.2 Software

Mission Planner (ArduPilot): Mission Planner is a comprehensive ground station application designed for use with the ArduPilot open-source autopilot project. It offers a range of tools and features for planning, configuring, and controlling autonomous vehicles such as UAVs, rovers, and boats [18].

Key Features of Mission Planner:

- Flight Planning: Allows users to create and modify flight plans with waypoint navigation, including take-off, landing, and various mission commands.
- Real-Time Monitoring: Provides real-time data monitoring and visualization of the vehicle's status, including telemetry, GPS position, and sensor data.
- Configuration and Tuning: Enables detailed configuration and fine-tuning of vehicle parameters to optimize performance.
- Simulation: Supports hardware-in-the-loop (HIL) and software-in-the-loop (SITL) simulations for testing missions and configurations without risking actual hardware.
- Firmware Management: Facilitates easy updating and installation of firmware for ArduPilot-based vehicles.
- Log Analysis: Includes tools for analyzing flight logs to diagnose issues and improve vehicle performance.

5.4 CALCULATION

To achieve sufficient agility and vertical climb performance, UAVs require a thrust output that significantly exceeds their total weight [19] (Table 5.4).

Thrust-to-Weight Ratio: Thrust produced by one propeller blade with 810 KV motor = 950 g. Total thrust produced by all propellers:

Table 5.4 Preliminary weight

Sr. no.	*Component name*	*Quantity*	*Weight*
1	Flight Controller	1	120 g
2	Electronics Speed Controller	6	168 g
3	BLDC Moter	6	320 g
4	LiPo Battery	1	680 g
5	Propellers	6	72 g
6	Receiver	1	17.6 g
7	UAV Frame	1	760 g
8	Telemetry Module	1	12 g
	Total	23	2149.6 g

$$\text{Total thrust} = \text{Number of propellers} \times \text{Thrust per propeller} \tag{5.1}$$

Substituting the values:

$$\text{Total thrust} = 6 \times 950 \text{ g} = 5700 \text{ g} \tag{5.2}$$

The total thrust produced is 5700 g. The formula for the thrust-to-weight ratio is:

$$\text{Thrust-to-weight ratio} = \frac{\text{Total thrust}}{\text{Total weight of the UAV}} \tag{5.3}$$

where total thrust is the total force produced by the motors (in newtons or grams, depending on units). Total weight is the total mass of the UAV, typically measured in grams or kilograms. Total thrust = 5700 g, total mass of UAV = 2149.6 g.

Substituting the values:

$$\text{Thrust-to-weight ratio} = \frac{5700}{2149.6} = 2.65 \tag{5.4}$$

The thrust-to-weight ratio is 2.65:1. The thrust-to-weight ratio between 1.9 and 3 enables UAVs to achieve high speeds and maneuver effectively in heavy wind conditions [19].

5.5 ISSUES FACED IN UAV DEVELOPMENT AND SOLUTIONS

Access the Flight Plan Screen: Navigate to the "Flight Plan" tab located on the left-hand side of the Mission Planner interface (Figure 5.3, Table 5.5).

Add a Waypoint:

- On the map displayed, right-click at the location to add a waypoint.
- A context menu will appear. Select "Add Waypoint" from the options.

Set Waypoint Parameters:

- A waypoint marker will appear on the map at the selected location.
- In the waypoint list below the map, configure specific parameters for the waypoint, such as:

Altitude: Set the desired altitude for the waypoint.

Loiter Time: Specify if the UAV should loiter (stay) at the waypoint for a certain period.

Table 5.5 Issues faced in UAV development and solutions

Issues	*Description*	*Resolution*
Battery Capacity Insufficiency	Battery capacity not meeting flight duration requirements	Upgrade to higher capacity batteries
Overheating ESCs/ motors	Excessive heat generation leading to ESC or motor failure	Select more efficient components
Accelerometer Misalignment	Misalignment of accelerometer axes affecting accuracy	Correct alignment through calibration
GPS Inaccuracy	Errors in GPS readings leading to navigation issues	Recalibrate GPS in open space
Magnetometer Interference	External magnetic fields affecting sensor calibration	Recalibrate in a location free from interference
ESC Calibration	ESCs not calibrated correctly leading to uneven motor performance	Calibrate ESCs to ensure uniform motor response
PID Tuning	Incorrect PID settings affecting stability and response	Tune PID settings for smoother take-off and flight

Figure 5.3 Adding waypoint to Mission Planner.

Waypoint Type: Choose from different waypoint types (e.g., regular waypoint, take-off, land).

Save and Upload the Mission:

- Once adding all the desired waypoints, review the waypoint list to ensure all parameters are set correctly.
- Click the "Write WPs" button to upload the waypoints to your UAV. This button is usually located at the bottom of the waypoint list.

Figure 5.4 UAV performing preplanned mission autonomous mission.

Verify the Mission: Click the "Read WPs" button to read back the waypoints from UAV and verify that they have been uploaded correctly.

Start the Mission:

- Navigate back to the "Actions" tab or the "Flight Data" screen.
- Arm UAV and initiate the mission. Select the "Auto" mode, which will start the UAV on the preplanned mission [18] (Figure 5.4).

5.6 RESULT

Time Efficiency in Agricultural Operations: In agriculture, the timely execution of tasks such as crop monitoring, pesticide spraying, and yield prediction is crucial to maximize yields and minimize losses. Our study demonstrates that autonomous UAVs significantly outperform manual flight in terms of time efficiency, which is particularly beneficial in agricultural settings. Autonomous UAVs: These systems can automatically navigate and perform tasks without human intervention, reducing the time required for operations such as crop health monitoring and precision spraying. Autonomous UAVs can cover larger areas more quickly and consistently, ensuring that critical agricultural tasks are completed in a timely manner [20].

Manual Flying: Manual operation of UAVs requires a skilled pilot to control the aircraft, which can be time-consuming and prone to human error. This method is slower and less efficient, especially in large agricultural areas.

Time Savings in Agricultural Operations experiment shows that autonomous UAVs consistently outperform manual flights over different distances. Flight tests were conducted under consistent weather conditions (*windspeed* < 5 *km*/h, temperature 32°C ± 5°C) without any payload (Table 5.6). Each test was repeated three times to ensure reliability of the data (Figure 5.5).

These results show that autonomous flights are faster than manual flights in all tested distances, with time savings ranging from 9.4% to 19%. The most significant improvement was observed at the shortest distance (50 m), which is typical for detailed crop monitoring or precision spraying tasks.

Efficiency Factors:

The data shows that autonomous control provides more efficient flight paths and potentially more stable flight characteristics, leading to reduced

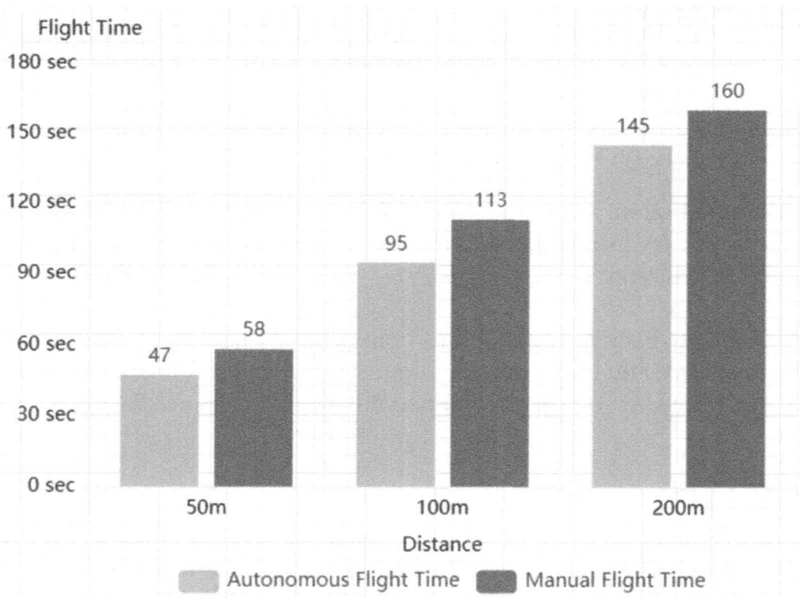

Figure 5.5 Graphical representation of manual and autonomous UAV operations flight time at different distances.

Table 5.6 Comparison of flight times between manual and autonomous UAV operations at different distances

No. of test	*Flight distance*	*Manual flight time*	*Autonomous flight time*	*Time saved*	*Improvement*
1	50 m	58 seconds	47 seconds	11 seconds	19%
2	100 m	113 seconds	95 seconds	18 seconds	15.9%
3	200 m	160 seconds	145 seconds	15 seconds	9.4%

flight times. This improvement in efficiency could be attributed to factors such as:

1. Precise Navigation and Path Planning: The autonomous system optimizes flight routes, ensuring that UAVs cover agricultural areas efficiently and consistently.
2. Consistent Speed Control: Autonomous UAVs maintain a steady speed throughout their flight, reducing variability and improving overall efficiency.
3. Optimized Acceleration and Deceleration Patterns: These systems can adjust acceleration and deceleration smoothly, minimizing unnecessary maneuvers and reducing flight times.
4. Reduced Human Error: Autonomous systems minimize the risk of human error, which can lead to inefficient flight paths or accidents, thereby ensuring safer and more reliable operations.
5. Better Battery Management: Autonomous UAVs can optimize flight patterns to reduce energy consumption, potentially extending flight times or allowing for more efficient battery use.
6. Enhanced Safety Features: Autonomous UAVs often include advanced safety features such as obstacle avoidance and emergency landing protocols, which reduce the risk of accidents and improve operational safety.

Implications for Agriculture:

- Crop Monitoring: Autonomous UAVs can rapidly survey large fields, detecting issues such as disease outbreaks or nutrient deficiencies much faster than manual methods. This allows farmers to respond promptly, reducing potential yield losses [20,21].
- Pesticide Spraying: Autonomous UAVs can spray pesticides more efficiently and accurately than manual methods, covering larger areas in less time. This not only saves time but also reduces the amount of pesticide needed, benefiting both the environment and the farmer's bottom line [21].
- Irrigation Management: Autonomous UAVs equipped with thermal imaging can rapidly assess soil moisture levels across entire fields, enabling farmers to optimize irrigation schedules and reduce water waste [20].
- Reduction in Pesticide Use: Autonomous UAVs can optimize pesticide application by precisely targeting affected areas, reducing chemical waste and environmental impact. This precision also leads to healthier crops and higher yields [21].

Increased Crop Health Monitoring Accuracy: By using multispectral or hyperspectral imaging, autonomous UAVs can accurately detect early signs of disease or stress in crops, allowing for timely interventions. Machine learning methods have been increasingly applied to UAV-collected data for improving crop health assessment and yield estimation.

Time Savings for Large-Scale Farms: The time efficiency of autonomous UAVs is particularly beneficial for large-scale farms, where manual methods would be impractical and time-consuming. This enables farmers to manage larger areas effectively, improving overall productivity.

Labor Cost Reduction: Autonomous UAVs reduce the need for manual labor in tasks like crop monitoring and spraying, which can significantly lower operational costs for farmers.

Table 5.7 presents a comparative analysis of manual versus autonomous flight performance across three different distances: 50, 100, and 200 m. The data clearly demonstrates the superior efficiency of autonomous flights over manual control. These results consistently show that autonomous flights are faster than manual flights across all tested distances. The time savings range

Figure 5.6 Mission Planner interface during UAV agricultural mission: real-time spraying and monitoring.

Table 5.7 Real-world testing conditions, limitations, and potential solution

Testing condition	*Limitation*	*Potential solution*
High Wind Speed	Reduced stability and accuracy	Implementation of control algorithms, use of wind-resistant UAV designs
GPS Signal Loss	Loss of navigation and positioning	Integration of alternative navigation systems (e.g., visual odometry (cameras), and LiDAR)
Obstacle Detection	Inaccurate detection of small obstacles	Integration of LiDAR sensors, use of advanced image processing techniques

from 9.4% to 19%, with the most significant improvement observed in the shortest distance (50 m). These findings underscore the potential benefits of autonomous UAV systems in applications where time efficiency is crucial, such as in search and rescue operations, delivery services, or time-sensitive data collection missions (Figure 5.6).

5.7 REGULATORY CHALLENGES, SAFETY, AND REAL-WORLD LIMITATIONS

The operation of autonomous UAVs is subject to a complex regulatory landscape, which varies across different countries and regions. In India, the DGCA has established guidelines and regulations for UAV operations, including requirements for registration, pilot licensing, and operational restrictions. These regulations aim to ensure the safe and responsible use of UAVs while promoting innovation and growth in the UAV industry [22].

Regulatory Challenges:

- Lack of Harmonization: The absence of globally harmonized regulations poses a significant challenge for UAV manufacturers and operators, as they must comply with different rules and requirements in each country where they operate [23].
- Evolving Regulatory Landscape: The rapid pace of technological advancements in UAV technology necessitates continuous updates and revisions to existing regulations to address emerging risks and challenges [23].
- Enforcement: Effective enforcement of UAV regulations is crucial for ensuring compliance and preventing misuse. However, limited resources and technological challenges can hinder enforcement efforts [23].

Safety Considerations:

- Collision Avoidance: Autonomous UAVs must be equipped with robust collision avoidance systems to prevent accidents and ensure the safety of people and property on the ground [16].
- Cybersecurity: Autonomous UAVs are vulnerable to cyberattacks, which could compromise their control systems and lead to safety incidents. Implementing robust cybersecurity measures is essential for protecting UAVs from cyber threats [16].

Privacy:

- The use of UAVs for surveillance and data collection raises privacy concerns, particularly in urban areas. Operators must adhere to strict privacy regulations and ethical guidelines to protect the privacy of individuals [24].

Real-World Limitations:

- Weather Dependency: The performance of autonomous UAVs can be significantly affected by adverse weather conditions such as rain, snow, and high winds.
- Battery Life: The limited battery life of UAVs restricts their operational range and endurance, limiting their applicability in certain scenarios.
- Public Acceptance: Public concerns about safety, privacy, and noise pollution can reduce the widespread adoption of autonomous UAVs [25].

5.8 SUMMARY

The study successfully demonstrates the design and implementation of an autonomous UAV system for precision agriculture. By integrating a Pixhawk 2.4.8 flight controller and Mission Planner software, the UAV showcased significant improvements in flight efficiency and task execution compared to manual operations.

Key features include:

- Real-Time Data Visualization: The interface provides live feedback on UAV position, flight path, and task progress, ensuring precise execution of spraying operations.
- Customizable Mission Planning: Farmers can define specific waypoints for targeted spraying, reducing chemical waste and improving resource efficiency.
- Monitoring Capabilities: The system allows for real-time assessment of field conditions, enabling immediate adjustments to spraying patterns or coverage areas.

Key Outcomes:

- Enhanced Efficiency: Autonomous flights outperformed manual flights across all tested distances (50, 100, 200 m), saving up to 19% in time. This efficiency is critical for time-sensitive agricultural tasks such as pesticide spraying and crop monitoring.
- Real-Time Agricultural Applications: The Mission Planner interface enabled precise execution of spraying missions with real-time monitoring capabilities. This ensures optimal resource utilization and better crop management.
- Versatility in Agriculture: The system's adaptability supports various agricultural applications, including crop health monitoring, irrigation management, and yield prediction.

However, challenges such as battery life limitations, power management, and obstacle detection highlight areas for future improvement.

REFERENCES

1. Gugan, G., and Haque, A., "Path Planning for Autonomous Drones: Challenges and Future Directions," *Drones*, Vol. 7, No. 3, 2023, p. 169. https://doi.org/10.3390/drones7030169.
2. Palossi, D., Zimmerman, N., Burrello, A., Conti, F., Müller, H., Gambardella, L. M., Benini, L., Giusti, A., and Guzzi, J., "Fully Onboard AI-Powered Human-Drone Pose Estimation on Ultralow-Power Autonomous Flying Nano-UAVs," *IEEE Internet of Things Journal*, Vol. 9, No. 3, 2022, pp. 1913–1929. https://doi.org/10.1109/JIOT.2021.3091643.
3. Mohsan, S. A. H., Khan, M. A., Noor, F., Ullah, I., and Alsharif, M. H., "Towards the Unmanned Aerial Vehicles (UAVs): A Comprehensive Review," *Drones*, Vol. 6, No. 6, 2022, p. 147. https://doi.org/10.3390/drones6060147.
4. Bhandari, S., Pathak, S., Poudel, R., Maskey, R., Shrestha, P., and Baidar, B., "Design and Development of Hexa-Copter for Environmental Research," *11th International Conference on ASIAN Community Knowledge Networks for the Economy, Society, Culture and Environmental Stability*, Kathmandu, Nepal, 2015.
5. Dalai, R. P., and Sabyasachi, S., "Future of Drone Technology in India," *Journal of Xi'an Shiyou University*, Vol. 15, 2019, p. 064X.
6. Mantha, R., "Development of an Unmanned Aerial Vehicle – A Tricopter," 2015. https://doi.org/10.13140/RG.2.1.2672.2729.
7. Lozano-Hernández, Y., Martínez de la Cruz, I., Gutiérrez-Frías, O., Lozada-Castillo, N., and Luviano-Juárez, A., "Design Procedure of a Low-Cost System for Energy Replenishment in a Quadrotor UAV through a Battery Exchange Mechanism," *Drones*, Vol. 7, No. 4, 2023, p. 270.
8. Oluwaseun, O., and Ehiagwina, F., "Design and Construction of Octocopter Agricultural Drone," *Iconic Research and Engineering Journals*, Vol. 5, 2022, pp. 62–74.
9. Priyambodo, T. K., Majid, A., and Shouran, Z. S. S., "Validation of Quad Tail-Sitter VTOL UAV Model in Fixed Wing Mode," *Journal of Robotics and Control (JRC)*, Vol. 4, No. 2, 2023, pp. 179–191.
10. Martinez, E., Garcia, M., and Rodriguez, P., "Aerodynamic Design Analysis for Long-Endurance UAVs," *AIAA Journal of Aircraft*, Vol. 28, No. 3, 2021, pp. 210–225.
11. Lee, K., Kim, S., and Park, H., "Propulsion System Design for Electric UAVs," *IEEE Transactions on Aerospace Systems*, Vol. 35, No. 2, 2022, pp. 180–195.
12. Khamsi, M., "Design and Analysis of an Unmanned Aerial Vehicle Capable of Carrying the Camera," *Universal Journal of Mechanical Engineering*, Vol. 7, 2019, pp. 87–96. https://doi.org/10.13189/ujme.2019.070302.
13. Nvss, S., Esakki, B., Yang, L.-J., Udayagiri, C., and Vepa, K. S., "Design and Development of Unibody Quadcopter Structure Using Optimization and Additive Manufacturing Techniques," *Designs*, Vol. 6, No. 1, 2022, p. 8. https://doi.org/10.3390/designs6010008.

14. Turanoguz, E., and Alemdaroğlu, H., "Design of a Medium Range Tactical UAV and Improvement of Its Performance by Using Winglets," *2015 International Conference on Unmanned Aircraft Systems (ICUAS)*, 2015, Denver, CO, pp. 1074–1083. https://doi.org/10.1109/ICUAS.2015.7152399.
15. Guebsi, R., Mami, S., and Chokmani, K., "Drones in Precision Agriculture: A Comprehensive Review of Applications, Technologies, and Challenges," *Drones*, Vol. 8, No. 11, 2024, p. 686. https://doi.org/10.3390/drones8110686.
16. Mohsan, S. A. H., Othman, N. Q. H., Yanlong, L., Alsharif, M., and Khan, M., "Unmanned Aerial Vehicles (UAVs): Practical Aspects, Applications, Open Challenges, Security Issues, and Future Trends," *Intelligent Service Robotics*, Vol. 16, 2023, pp. 109–137. https://doi.org/10.1007/s11370-022-00452-4.
17. Kim, S.-G., Lee, E., Hong, I.-P., and Yook, J.-G., "Review of Intentional Electromagnetic Interference on UAV Sensor Modules and Experimental Study," *Sensors*, Vol. 22, 2022, p. 2384. https://doi.org/10.3390/s22062384.
18. Chintanadilok, J., Patel, S., Zhuang, Y., and Singh, A., "Mission Planner: An Open-Source Alternative to Commercial Flight Planning Software for Unmanned Aerial Systems," 2022.
19. Octonius, R., and Ruseno, N., "Endurance Estimation in Hovering Flight Based on Battery Power Requested on Quadcopter UAV," *Angkasa: Jurnal Ilmiah Bidang Teknologi*, Vol. 14, 2022, 119–132. https://doi.org/10.28989/angkasa.v14i2.1226.
20. Tsouros, D. C., Bibi, S., and Sarigiannidis, P. G., "A Review on UAV-Based Applications for Precision Agriculture," *Information*, Vol. 10, No. 11, 2019, p. 349. https://doi.org/10.3390/info10110349.
21. Liu, Z., and Li, J., "Application of Unmanned Aerial Vehicles in Precision Agriculture," *Agriculture*, Vol. 13, No. 7, 2023, p. 1375. https://doi.org/10.3390/agriculture13071375.
22. Mehrotra, K., "Drone Regulations: The Indian Scenario," *SSRN*, May 2024. https://ssrn.com/abstract=4817307.
23. Rajagopalan, R., and Krishna, R., "India's Drone Policy: Domestic and Global Imperatives," *ICAO Scientific Review: Analytics and Management Research*, Vol. 1, 2019, pp. 53–68. https://www.informingscience.org/Publications/4382.
24. Makam, G., "Navigating the Skies: Assessing the Regulatory Landscape and Future Implications of Drone Law in India," *SSRN*, Jun 2023. https://ssrn.com/abstract=4480534.
25. Mane, M., Sonawwanay, P., Solanki, M., Manu, M. V., Sonawwanay, P. D., Solanki, M., and Patel, V., "A Comprehensive Review on Advancements in Noise Reduction for Unmanned Aerial Vehicles (UAVs)," *Journal of Vibration Engineering & Technologies*, Vol. 12, No. Suppl 2, 2024, pp. 1375–1397. https://doi.org/10.1007/s42417-024-01480-3.

Chapter 6

Advanced crop monitoring using unmanned aerial vehicle with Internet of Things and cloud computing

Sumit Kumar Jindal and Muthke Vivek

6.1 INTRODUCTION

This chapter discusses how unmanned aerial vehicles (UAVs), Internet of Things (IoT) devices and cloud computing together can help farmers to remotely monitor and optimize their agricultural yield. It also includes UAV-based Received Signal Strength Indicator (RSSI) scanning that can identify locations with maximum connectivity and least interference. Agriculture has been one of the oldest and most important industries and is in urgent need of transformation. Food shortage is expected to become a major problem due to the growing world population and could only be met by efficient agricultural practices. Challenges like climate change and unpredictable weather have made it more difficult to predict crop behaviour and make right decisions that provide highly nutritious food. Therefore, the chapter delves into the importance of real-time monitoring and making strategic decisions that optimize agricultural results. UAVs equipped with high-resolution cameras and sensors have become a very necessary tool for precision agriculture. These enable surveying large areas of agricultural lands that were previously difficult to cover. IoT devices can help the farmers to attain vital information like soil moisture, soil pH, weather, chlorophyll content, pest and disease symptoms. The raw data attained from the sensors can be huge in size, which brings in new complications like worsened latency, bandwidth usage, and response times. Solutions like edge computing have been discussed that offload processing from a centralized server to a device closer to the sensors. These processed data can be fed to Artificial Intelligence (AI)-driven algorithms that analyse the patterns and serve as tools for decision-making. In spite of having such great potential, one of the major setbacks could be establishing stable communication among the IoT devices. To ensure the viability of the solution in expansive farming, transmitting data over vast lands and ensuring the strategic placement of IoT base stations is very necessary. One effective way of tackling this issue can be scanning for signal strength all across the fields by measuring RSSI

DOI: 10.1201/9781003589730-6

values using an UAV by applying clustering algorithms that can analyse the raw signal strength data and identify locations with the least interference for ideal positioning of the devices.

6.1.1 History

Military and Industrial Origins (Before 2000s):

- UAVs (drones) were first used for military purpose, where they were used for surveillance, reconnaissance, and target acquisition.
- Later in the 1990s drones were used to serve industrial applications such as infrastructure inspection and disaster monitoring.
- The early drones were expensive, huge and required specialized operators, making their applications limited to military and industrial use cases.

Transition to Agricultural Applications (Early 2000s):

- Around the early 2000s, researchers and agricultural scientists started experimenting with UAVs for farming applications.
- One of the early applications of drones apart from military and industrial applications was aerial imaging of fields for crop health monitoring using RGB and near-infrared (NIR).
- Companies like Yamaha from Japan are one of the early adopters and played an important role with their inventions like automated spraying systems in rice fields.

Challenges and Limitations:

- Regulatory Hurdles: Many countries had restricted UAV flights in rural areas with their airspace regulations.
- Short Battery Life: Early UAVs used low-capacity batteries due to payload constraints and low energy density, limiting their flight time to 15–30 minutes.
- Data Processing Limitations: With no advanced software that could analyse the acquired data in real time, decision-making was based on human interpretation which was slow and biased sometimes.
- Lack of Automation: Precision agriculture for large-scale applications was impractical as drones had to be manually operated without the automated Global Positioning System (GPS)-based systems.
- High Costs: Commercial drones and imaging equipment are expensive, therefore only in the reach of large research institutions or scale farms.

Key Breakthroughs Before 2010:

- By the late 2000s, energy-dense batteries like lithium-polymer (LiPo) batteries, open-source flight controllers and advancements in GPS-based navigation have made UAVs more affordable and accessible.
- The potential of UAVs for precision agriculture was demonstrated in research projects across Japan, the U.S. and Europe, spurring further investment.
- Multispectral and thermal cameras enabled the early identification of plant stress and irrigation problems.

6.1.2 IoT in smart farming: real-time monitoring and data collection

IoT has played a crucial role in transforming agriculture by enabling remote monitoring and automating tasks, making it easy to practice large-scale agriculture with smart resource management. Various sensors in an interconnected network collect data on essential plant parameters like soil moisture, temperature, humidity, and other environmental conditions. This data is wirelessly transferred to a centralized cloud-based system [1], where it is processed and structured for farmers to access from anywhere. Farmers can rely on data-driven predictions to strategically make decisions for spraying pest controls, fertilizers, and irrigation timings. Additionally, the system could be automated by having predefined thresholds, which, when reached, can trigger particular agricultural tasks that were otherwise carried out manually. This decreases dependency on labour, making it cheaper and more efficient with timely interventions.

6.1.3 Role of UAVs in crop monitoring: aerial surveillance and analysis

Aerial imaging and real-time surveillance of agricultural lands using UAVs have transformed the farming sector. The ability of humans to perceive only the visible light spectrum results in a significant loss of imagery data across other spectral ranges, like the infrared spectrum, which detects changes in temperature due to water stress, nutrient imbalance and underlying diseases. Also, traditional manual checks are often time-consuming, inefficient and reliant on the observer's inspection. All these factors make a drone equipped with high-resolution and infrared cameras fundamentally a better solution for analysing image-based data. Laser-based scanning technologies like Light Detection and Ranging (LiDAR) can be used to 3D map the terrain and canopy to understand crop density, plant height and biomass.

Beyond monitoring, UAVs coupled with spraying systems can be employed for spraying pesticides and fertilizers [2], which again decreases labour and effectively reduces the overall cost of maintaining farmland.

6.1.4 RSSI-based network optimization: ensuring seamless communication in IoT agriculture

RSSI is a metric used to determine signal power received from a radio-frequency (RF) client at a particular location. RSSI values are usually represented in negative numbers in dBm and closer the value to zero, better the signal strength. Signal power in a given area is dependent on various factors such as distance from the source, antenna orientation, obstacles and interference. Deploying IoT devices in the locations with high RSSI ensures stable network connection and transmission reliability [3]. Geographical coordinates with high RSSI can be located using a GPS module and micro-controller with desired network capabilities. This module on a UAV can cover large fields that were otherwise difficult to inspect. With GPS coordinates and their corresponding RSSI values, zones with optimal signal strength can be identified using clustering algorithms that groups and categorizes data points.

6.2 PROPOSED SOLUTION

This proposal aims to make farming more efficient by using drones, IoT sensors, and RSSI-based analysis for real-time crop monitoring [3]. This enhances resource management, minimizes expenses and improves our ability to predict harvests. Using drones to collect data and arranging IoT sensors based on signal strength patterns, the system ensures stable communication between field sensors and remote monitoring platforms.

System Overview:

As shown in Figure 6.1, the quadcopter mainly features a flight controller, motors, Raspberry Pi and a system to communicate to the cloud server.

1. Flight Controller: Flight control is the most important component in a UAV that is responsible to maintain balance and navigate. To do this, it incorporates various sensors such as accelerometer, gyroscope, GPS, magnetometer and barometer. The gyroscope and accelerometer determine the linear acceleration, orientation with respect to pitch, roll and yaw axes; the magnetometer provides orientation with respect to earth's magnetic field which is crucial to know the heading direction of the drone. And the altitude of the drone can be estimated using a barometer which calculates atmospheric pressure. The flight control uses control algorithms with

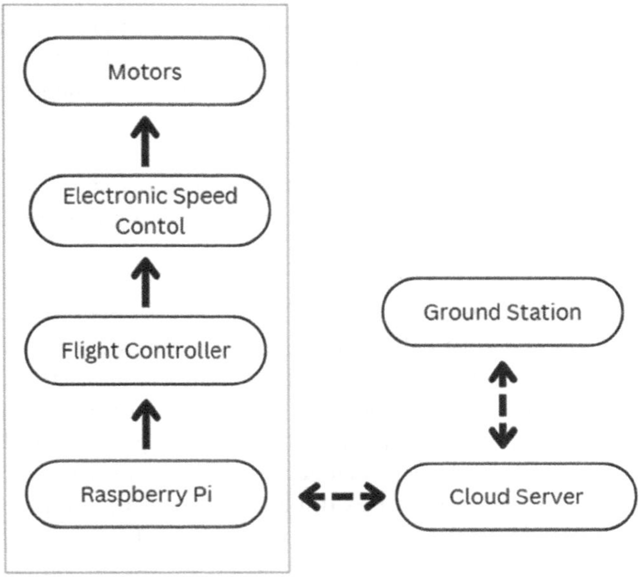

Figure 6.1 System architecture of the quadcopter.

the sensor data to control the speed of the motors to stabilize the vehicle. Kalman filter, Proportional-Integral-Derivative (PID) control and complementary filter are some of the popular control algorithms used. For example, if the drone begins to tilt in a particular direction, the sensor data captures it and the control algorithm signals to increase the motor speed to compensate for the tilt and bring the UAV back to level. In autonomous drones, the flight control includes a pre-programmed flight plan using GPS waypoints. The controller may also be programmed to avoid obstacles and replan to match GPS waypoints.

2. Motors: Motors in a UAV drive the propellers, providing thrust to take off and steer through air. UAVs are usually equipped with four sets of motors that vary rotational speeds in accordance with the data received from the flight controller. The most common motors used are DC brushless motors due to their efficiency and reliability compared to traditional brushed motors that have power losses due to internal friction generated while rotating. For agricultural applications where drones have relatively high payload requirements, larger propellers and lower pitch are used to generate more thrust. But this in turn produces more drag and air resistance that can be tackled using high toque motors.

3. Raspberry Pi (Onboard Computing): The Raspberry Pi serves as the onboard computer for the UAV, managing data collection from the various sensors and handling communication with the base station. The Raspberry Pi is connected to several key components:
   - Sensors: It receives real-time data from environmental sensors like temperature, humidity, soil moisture and cameras (multispectral or thermal) that are crucial for crop monitoring.
   - GPS Module: It receives geolocation data to guide the UAV along the flight path.
   - Communication Modules: It houses the Wi-Fi, LoRa or 5G communication modules to transmit data to the base station.
4. UAVs Equipped with IoT Sensors: Plant health and growth are determined by nutrients and its surrounding environment and in turn the changes in plants have subtle effects on the environment. These effects can be monitored to determine various plant parameters like turgor pressure, photosynthesis efficiency, pest resistance, growth rate and flowering and fruiting. For example, oxygen production declines and carbon dioxide levels rise locally for an unwell plant. High humidity levels promote fungal growth whereas low humidity increases transpiration, resulting in dehydration. Low temperatures slow down metabolism and high temperatures lead to heat stress which can influence flowering and fruiting cycles. UAVs outfitted with IoT sensors can play a crucial role in maintaining conditions for optimal stomatal functioning.
5. RSSI-Based Network Optimization: To ensure efficient data transmission between the IoT devices, the system uses an RSSI-based network optimization strategy. It identifies the best location to deploy the IoT devices, ensuring a low interference communication. UAVs are equipped with a module that can scan for RSSI all over the fields. The RSSI data is then processed through a clustering algorithm that divides the fields into zones depending on the signal strength. This data-driven approach for deploying IoT base stations ensures stable, reliable and high-speed communication. This also reduces packet loss that occurs due to network interference and poor signal strengths.
6. Cloud-Based Monitoring Platform: Unlike traditional computing that works on the local hardware, cloud computing is an online computing service that handles tasks like servers, database, storage, processing and networking. This is an inexpensive solution that enables remote access and is maintenance free. It also allows data syncing in real time through various edge devices to make decisions based on latest information and as it offers backup and data security it makes the data recoverable even when the whole system fails. Cloud computing can be categorized to three types based on deployment model: public, private, hybrid and community cloud.

Public Cloud – Public cloud refers to cloud that is operated and provided by publicly available parties like AWS and Google Cloud. It offers limited customization and control over the infrastructure.

Private Cloud – Private cloud is exclusively controlled by a single organization and offers more security when compared to a public cloud. Hybrid cloud is a mix of both private and public clouds.

Hybrid Cloud – Hybrid cloud includes features of both public and private clouding, making it more flexible and can be a cost-effective solution if only a portion of your data needs high security.

Community Cloud – A community cloud shares the platform with organizations and individuals of similar interests. This could split the expenses and while being optimized for specific needs.

While cloud solutions have many advantages, they also present certain challenges. These challenges include:

Downtime and Reliability – Though cloud providers are highly reliable, sometimes technical issues and maintenance can cause interruptions.

Vendor Lock-in – Once signed up for a particular cloud service it can be very difficult to transfer data and migrate to another service provider. Therefore, it's very important to choose the right vendor by careful evaluation.

6.3 ARCHITECTURE OF THE SYSTEM

The system consists of four different layers: Perception, Network, Data Processing and Application that possess distinct functionalities. All these layers work together seamlessly to suggest ideal decisions that are required for plant growth.

The system follows a layered architecture to organize data flow, ensuring productive data collection, processing, analysis and presentation. These layers work together seamlessly, offering a holistic solution for accurate agriculture. The architecture exists in four layers: Perception, Network, Data Processing and Application.

6.3.1 Perception layer (IoT sensors and UAVs)

This is the data acquisition layer, where sensors collect environmental information from the land field. The UAVs play a key role in gathering high-resolution data over large areas, providing both spatial and temporal context.

Components:

- Soil Moisture – Tracks water content in the soil which determines watering needs [4]. The sensor provides a voltage output that can be read by the microcontroller to know how moist or dry the soil is. A range of soil moisture sensors exist with different working principles.
- Resistance Variation-Based Soil Moisture Detection – This type includes two electrodes that are placed in the soil. As the soil gets wetter the ions in the water decrease the resistance between the electrodes, resulting in high current. Resistance-based detectors are usually cheap but corrode over time, leading to inaccuracies.
- Capacitive Soil Moisture Sensor – Works like a regular capacitor where the plates are replaced by electrodes and the dielectric material by the soil. As the water content increases in the soil the dielectric constant increases. Capacitive-based soil moisture detectors are expensive and more durable compared to resistive soil moisture sensors.
- Time Domain Reflectometry (TDR) Sensor – A probe inserted in the soil is stimulated with an electromagnetic pulse and the time taken for the pulse to travel along the probe is calculated. This time varies with variation of moisture in the soil. The wetter the soil, the slower the travelling speed of the pulse. TDR sensors are highly accurate and reliable but are more expensive and complex.

Capacitive-based sensors could be ideal for agricultural applications where accuracy and long-term reliability are the priorities.

- Temperature Sensor – Temperature sensors play a major role in maintaining optimal conditions for plant growth, germination and pest control [5]. Various temperature sensors are available in the market; some of them include:
    - Thermocouples – This technology uses two wires of different metals that are joined at an end and variation in temperature at this junction produces a tiny voltage which when amplified can determine the temperature. Thermocouples can measure a wide range of temperatures and have quick response but are sensitive to other external factors like humidity.
    - RTDs (Resistance Temperature Detectors) – These sensors have relatively slow response times but are more accurate and less sensitive to environmental noise. They work on measuring change in resistance which is a function of temperature.
    - Thermistors – Another type of temperature sensor that operates on a similar principle to RTDs but use ceramic instead of metal, making them more sensitive and cost-effective. The only downside of thermistors is that they have limited temperature range compared to the RTDs and thermocouples.

- Infrared Temperature Sensors – Infrared sensors are non-contact-based sensors that detect infrared radiation emitted by objects to measure the surface temperature. These can be a quick and efficient solution but are affected by smoke and dust.

RTDs and thermistor or semiconductor sensors are commonly preferred for temperature monitoring in agricultural settings due to their reliability and accuracy.

- Humidity Sensors: Selecting the right humidity sensors to monitor moisture levels in the air is very crucial to optimize irrigation timings. Here are some humidity sensors available, each one having their unique set of advantages.
  - Capacitive Humidity Sensors – These are ideal for greenhouse monitoring and indoor farming due to their low maintenance and accuracy. The sensor consists of hygroscopic material and the change in capacitance is used to measure humidity levels.
  - Resistive Humidity Sensors – Just like the capacitive humidity sensor, this sensor uses a hygroscopic material but relies on variation in resistance. As humidity increases, electrical resistance in the sensor decreases and after easy calibrations the sensor can measure humidity levels. Being cheap and simple to use, resistive humidity sensors are best suited for open field farming.
  - Dew Point Hygrometers – Dew Point Hygrometers are extremely accurate and can be used in extreme conditions. They work by cooling a mirror or the sensor surface and as the temperature reaches dew point the moisture in the air starts to condense on the surface. This condensate is detected either by optical methods or by measuring thermal properties of the surface. Humidity can be hence calculated as dew point and humidity are directly proportional. These are mainly used in high-precision agriculture like greenhouses and vineyards.
- Multispectral and Thermal Cameras: Plants under stress are warmer and emit infrared radiations that are not possible to monitor with naked eye but can be detected using thermal cameras [6,7]. The stress could be due to multiple factors like water stress and underlying diseases. Being able to detect the variations in infrared radiation can help farmers to optimize irrigation and spray localized insecticides. Normalized Difference Vegetation Index (NDVI) is an important vegetation index that helps in assessing plant health using multispectral imagery. Healthy plants reflect large amounts of NIR light and absorb visible red light. NDVI is calculated from these parameters and has range from –1 to +1. Negative values indicate that it's a non-vegetated area with water bodies. Positive values nearer to +1 indicate dense vegetation and nearer to zero represent deserted and almost barren land.

- GPS Module: GPS modules are electronic devices that are used to map the collected data with the geographic coordinates in a farmland without which the data would be irrelevant. It's also used to automate drone using predefined flight paths.

The GPS system works on a principle known as trilateration. In trilateration, a GPS module is connected to minimum four satellites and each of the satellites transmits continuous signals. The module calculates the time taken to receive signal and triangulates the exact location in all three axes (latitude, longitude and altitude).

Selecting a GPS module can be challenging and it's important to consider precision, reliability and budget.

RTK GPS – RTK GPS stands for Real-Time Kinematic and is best for high-precision applications like NDVI mapping and performing detailed crop analysis. It provides centimetre-level precision but is expensive.

PPK GPS – PPK GPS stand for Post-Processed Kinematic and can be a cheaper solution providing similar accuracy to RTK. The only drawback is that unlike RTK it is not capable of performing real-time corrections during flight.

DGPS – DGPS stands for Differential GPS and is more accurate than basic consumer-grade GPS but far less precise when compared to PPK and RTK GPS. These are best suited for most of the agricultural tasks like navigation, monitoring and broad surveys.

- Wi-Fi and LoRa-Based Communication Modules: These modules are crucial to enable communication between the UAVs, edge devices and base stations. It facilitates smart agriculture for remote monitoring and automated control. Wi-Fi and LoRa are two most commonly used technologies that are used for short- and long-range communications respectively.

LoRa – LoRa sensor can be used to transfer sensor data like soil moisture, humidity and temperature to nearby LoRa gateway.

Wi-Fi – Data from LoRa gateway is sent to the cloud for further processing and this data is uploaded to cloud-based dashboards that can be accessed by farmers to make decisions.

6.3.2 Network layer (RSSI-based optimization and base stations)

The Network Layer is responsible for the placement of base stations in optimum locations to enhance network coverage and decrease interference by clustering RSSI at different geographic coordinates in the farmland.

The network layers have three key components:

1. RSSI Measurement Algorithms
2. Clustering Algorithms (*K*-means, DBSCAN, etc.)

1. RSSI Measurement Algorithms: The RSSI measurements determine the signal strength between devices that can be used to identify regions with weak signals that require additional base stations to improve the signal quality. This helps in reducing packet loss and communication latency in the network. Wi-Fi modules (ESP8266/ESP32), LoRa Modules (RFM95W) and particularly for mesh networking Zigbee modules (XBee Series 2, CC2530) can be used to determine RSSI values of a particular location. Here, we use a UAV to scan for RSSI measurements and its corresponding GPS locations all over the field.
2. Clustering Algorithms (*K*-means, DBSCAN, etc.): Clustering Algorithms analyse and group data with similar characteristics. Here, RSSI data is grouped into different zones to improve the signal strength (refer Figure 6.2). Mentioned below are some key clustering techniques:
   - *K*-Mean Clustering: *K*-Mean Clustering is a popular algorithm used in machine learning, image processing and

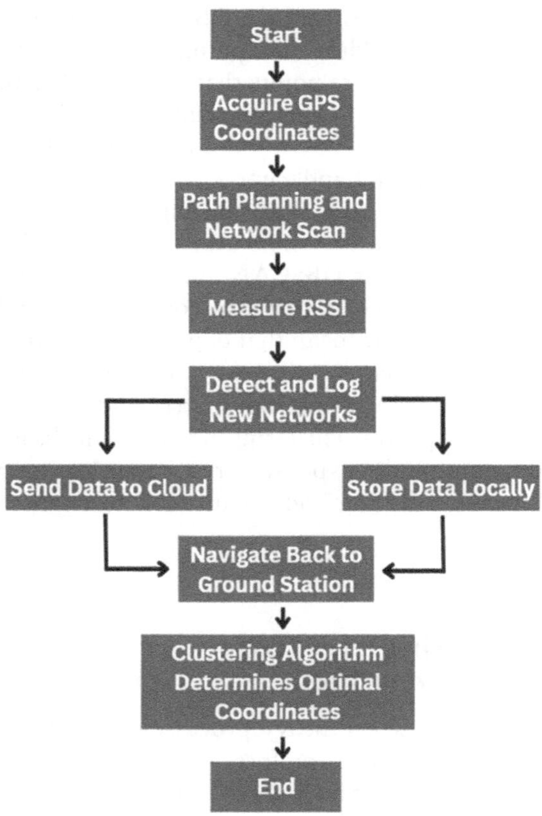

Figure 6.2 Block diagram illustrating the RSSI-based navigation algorithm.

network optimization. The number of clusters formed is determined by *K* [8]. First, *K* random centroids are selected on the cartesian plane and the data points nearest (Euclidean distance) to these centroids are grouped. The mean of these grouped data points is taken and the new centroid is relocated to the mean and clusters are formed again on the basis of Euclidean distance. This is repeated until centroids no longer change. Finally, the algorithm outputs *K* clusters with similar characteristics.

Note – Clustering data that contains noise, outliers and very irregular clusters tends to fail using *K*-mean clustering.

- DBSCAN: DBSCAN is a density-based clustering algorithm that groups data points based on how dense or close they are. Before understanding DBSCAN certain parameters and naming should be discussed. Neighbourhood radius (*E*) and minimum number of points required to form a dense region (*M*) are user-defined parameters. Core point is a point with at least *M* neighbours within *E* radius. Border point is defined as a data point that is within *E* radius of a core point but has less than *M* neighbours. To perform DBSCAN start expanding from a core point by adding data points within *E* radius. Repeat this until there no more points that can be added and perform the same operation on the next unvisited point. Unlike *K*-mean clustering [8,9], DBSCAN doesn't require defining number of clusters and can handle noise.
- Spectral Clustering: Spectral clustering is another way of clustering that can handle arbitrary cluster shapes which are very difficult for DBSCAN such as concentric circular and crescent clusters. It relies on concepts of similarities unlike DBSCAN and *K*-mean that depend on Euclidean distances. It does this by representing the data into a graphical representation and using Laplacian matrix and eigen vectors to form clusters. Spectral clustering has its own disadvantages. It is computationally expensive, requires selection of value *K* like *K*-mean clustering and it's difficult to choose the similarity matrix.

6.3.3 Data processing layer (edge and cloud computing)

Data processing layer is where heavy processing and data analytics take place. This layer is responsible for forecasting agricultural outcome and making AI-driven decisions. It includes both edge and cloud computing.

- Edge Computing Devices: Edge computing refers to processing the data near the data source instead of only relying on centralized cloud servers. This reduces latency, bandwidth usage and increases system efficiency. It also allows to scale easily without overburdening the cloud infrastructure. In agricultural context that incorporates thousands of devices it plays a crucial role in optimizing resources and power efficiency. Choosing a right edge device depends on the tasks that are being performed locally. Some of the most popular edge devices have been listed below.
- Raspberry Pi: An affordable and versatile single board computer that has wide community support. It is ideal for general purpose computing and can run local data analytics that can be used for smart irrigation systems. It is used by hobbyists, for prototyping and sometimes even in industrial environments.
- NVIDIA Jetson (e.g., Jetson Nano, Jetson Xavier): It's a high-performance edge computing device that has GPU capabilities which makes it optimized for image processing and AI tasks. It can be used for advanced use cases like real-time crop monitoring using UAVs for multispectral plant disease detection [10]. The only drawbacks are that it is expensive and has high power consumption.
- Arduino: Arduino is a basic, low-cost open-source microcontroller that can gather sensor data and control actuators. It's not suitable for complex data analysis and AI tasks but can perform simple tasks like automating irrigation and fertilization when sensor data reaches a specific threshold.
- Google Coral: Google Coral is ideal for performing simple machine learning tasks and energy efficiency is the priority. It is optimized to run TensorFlow Lite which is an open-sourced framework from Google that is used for training and deploying machine learning models.
- Cloud Storage and Processing Servers: Cloud storage and processing involves storing, processing and managing data in online servers that are housed in large data centres instead of a local system. This makes it flexible and accessible from anywhere at any time. The cloud server also provides high-end processing for data analytics and machine learning tasks that would be very expensive to implement on local hardware otherwise. This includes batch processing, real-time analytics and complex simulations. Cloud storage companies offer different services based on the customer needs. These include:
- Object Storage: Services like Amazon S3, Google Cloud Storage, Microsoft Azure Blob Storage fall into this category and is best for storing unstructured files like videos and images. In the agricultural context, raw data from various sensors, drones and weather stations with file types .csv, .json, .xml, .parquet can be stored with this technology.

- Block Storage: Amazon EBS, Google Persistent Disks, Azure Disk Storage are some examples of block storage. These provide more control over storage volumes and direct access to the data blocks which makes it ideal for applications with low latency and high performance.
- File Storage: Google Cloud Filestore, Azure File Storage, Amazon EFS are some examples of files storage. It is ideal for applications that require a collaboration and shared file for web application and content management systems.

Depending on computational workloads cloud processing platforms need to be cleverly evaluated for optimal performance. If the tasks are related to simple data storage, IoT communication and lightweight processing, Google Cloud Compute Engine, AWS EC2, Lambda and Microsoft Azure Virtual Machines are the way to go but for complex AI-based workloads and big data analytics GPU/TPU processing is required which are provided by NVIDIA DGX Cloud, Google Cloud AI (TPUs) and AWS AI & SageMaker.

- WebSockets for Real-Time Data Streaming: WebSocket is communication protocol that is used for real-time applications and enables two-way communication between client and the server. Unlike traditional HTTP that requires multiple connections for transferring data, WebSockets offer low-latency persistent connection. Farmers can receiver instant alerts when crops experience critical environmental changes. WebSockets can also be used in the case of continuous streaming of UAV data for real-time analysis and navigation adjustments.

6.3.4 Application layer (visualization and decision support)

The application layer is important for visualizing the data collected and providing actionable insights to aid in decision-making. It serves as an interface for the users. The application layer has three main components.

Components:

- AI-Driven Analytics for Predictive Insights: Machine learning models generate insights such as yield predictions, early disease detection, and optimized resource usage by analysing the collected data.
- Mobile and Web Applications with Dashboard Visualization: These apps present real-time data on crop health, environmental conditions and the status of automated methods like irrigation. Dashboards provide a clear, easy-to-comprehend representation of data, helping users make informed decisions tireless.

- Alert Systems: When anomalies are detected, such as soil moisture falling below a certain threshold or signs of crop disease, the system sends alerts and notifications. This allows farmers to take corrective actions promptly.

6.4 IMPLEMENTATION STRATEGY

This section covers a detailed breakdown on the implementation strategy that can be adopted to deploy a system that ensures efficiency, optimum resource utilization and real-time monitoring. It describes thoroughly UAV deployment, network optimization using RSSI-based base station placement, data collection and cloud-based data analysis.

6.4.1 UAV deployment and route planning

To ensure efficient and structured data collection, UAVs are programmed to autonomously follow specific flight routes that depend on predefined GPS waypoints. This path is designed using a Ground Control Station (GCS) software before take-off. The drone adjusts thrust, direction, speed and orientation to meet the waypoints. This also reduces redundant data captured by overlapping flight paths and thereby extending battery life. It can also be programmed to avoid obstacles using various proximity sensors and redefine flight paths. Additionally, geofencing technology is integrated to prevent the drone from flying outside the desired boundaries, ensuring safety and regulatory compliances. For most of the fields, grid-based sweeping works but to identify the best route plans algorithms like Genetic Algorithm (GA) Optimization can be used. The algorithm generates random routes and evaluates time of flight, battery consumption, obstacles and then the optimal paths are either combined, mutated or adopted for efficient routing [11].

Data collection – To collect high-resolution farm data, UAVs utilize advanced imaging systems, such as multispectral and hyperspectral cameras. These sensors capture information on early symptoms of plant diseases, water stress and nutrient deficiencies that are invisible to the human eye. The sensors in the field complement the UAV assessment capabilities and enhance irrigation and pest control strategies. Additionally, GPS modules help correlate aerial insights with specific field locations to take targeted measures. All the collected data can either be stored on board or transmitted via wireless protocols like Wi-Fi, LoRa and 4G/5G depending on the network connectivity. To minimize the bandwidth requirements the data from the UAVs can be compressed. Data is either transferred to ground station or cloud. In many cases, cloud storage is preferred as it can be scaled easily, making it ideal for large data volumes.

6.4.2 RSSI-based base station placement

The need for RSSI-based base station placement and it's working has been discussed in Section 6.3.2 "Network Layer". Now, we will proceed to discuss on testing and validation of the proposed setup to ensure it meets the desired performance criteria. This includes both simulation and real-world testing.

1. Simulation of Base Station Placement: The process includes shaping the environment by containing terrestrial details like cliffs, buildings, thickets and weather environments like rain, snow and fog that could affect the outputs. Use propagation models (free-space path loss, urban path loss, etc.) to estimate signal strength and behaviour in the modelled environment. Then run the simulation to assess whether the base stations provide the desired coverage with minimum overlapping and interference. During simulation measure performance metrics such as Signal-to-Noise Ratio (SNR), Data Throughput and Coverage Area. Also, test the system limits by introducing more devices or users and then make refinements to further test it in the real world.
2. Field Testing (Real-World Testing): Field Testing involves deploying mobile units, such as drones and IoT devices, in the field equipped with RSSI sensors and GPS tracking. Set up a controlled environment where devices systematically move to predefined paths and measure signal strength so that it's easier to verify and align with the simulated results. Measure SNR to ensure the signal quality is sufficient for reliable communication. SNR is the ratio of signal strength to the noise and is measured in decibels (dB). Understand and analyse the results to fine-tune the height, orientation and power of each base station to achieve better coverage. Other performance metrics include coverage area and data throughput. Data throughput refers to the rate at which data can be successfully transferred across the network. To get more practical results use tools that can simulate data traffic like video streaming or file transfers and measure throughput across the network.

6.4.3 Data collection and transmission

IoT sensors deployed across the field along with UAVs collect crucial data that can be used to make predictions and manage agricultural resources. Table 6.1 summarizes the purpose of each sensor used in these settlings.

Sensor data collection can be divided into three broad categories:

- Continuous Monitoring – The sensor systems collect data without any interruption. This is often used in dynamic environments where inputs rapidly change. Continuous monitoring provides high-resolution data

Table 6.1 Types of sensors for crop monitoring

Sensor type	*Purpose*	*Data collected*
Soil Moisture	Optimize irrigation timings	Soil moisture levels
Temperature and Humidity Sensor	Monitor climate changes	Environmental temperature and humidity
Soil pH Sensor	Optimize fertilizer usage	Acidity/alkalinity of soil
CO_2 and Gas Sensors	Detect plant disease and monitor respiration	Air quality, gas concentration
Thermal Camera	Identify irrigation issues and underlying stress	Temperature variations in crops
Multispectral Camera	Monitor crop stress	Vegetation health (NDVI)
LiDAR Sensor	Measure plant growth and canopy	Terrain mapping

Table 6.2 Types of storage devices

Device	*Purpose*	*Data logged*
SD Cards and Solid-State Drives (in UAV Storages)	Offline logging in areas with weak connectivity	High-resolution footages and images
Cloud Databases	Long-term storage and analytics	Processed and raw data from IoT devices and UAVs
Microcontroller Storage (Arduino, ESP8266 etc.)	Short-term and limited data sets	Sensor readings
Local Servers (PC and On-Premises Servers)	Serves in case of cloud failure and downtime	Real-time data backup

and allows immediate decision-making. It requires massive data storage and consumes high power.

- Time-Based Sampling – Sensors are programmed to collect data in fixed intervals of time. This makes the system energy efficient and simplifies data processing but is inefficient for dynamic systems as it can miss important events.
- Event-Based Sampling – This collects data whenever there is a trigger. For example, collecting humidity data only when the temperature sensors measure temperature above 30°C. Though the sensors could activate due to false triggers, most of the time it ensures efficient data usage as measurements are taken only when necessary.

Before further transmission, the collected data needs to be logged. Table 6.2 provides the purpose of various data logging solutions.

From Figures 6.3 and 6.4, LoRa is optimal for long-range and lower-power wireless data transmission like simple data from IoT sensors due to its low data transfer speeds (Figure 6.5). Let's consider a TDR sensor for soil moisture that needs to transmit data via LoRa. The soil moisture sensor

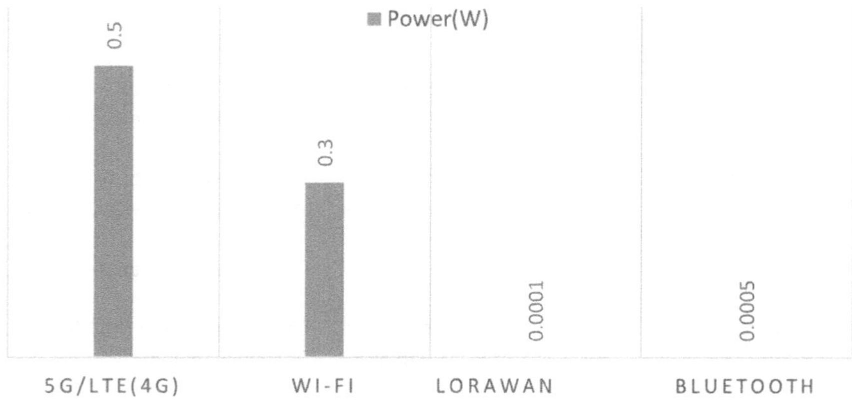

Figure 6.3 Graph representing power consumption of different wireless technologies.

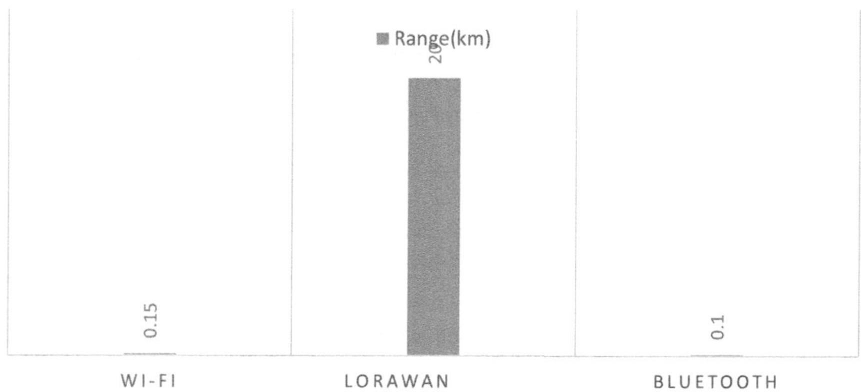

Figure 6.4 Graph representing range of different wireless technologies.

is considered as LoRa node and multiple nodes across the farmland transmit data to the LoRa gateway from where it is sent to the cloud. Since the LoRa devices themselves don't have capabilities to transmit to the cloud, a LoRa gateway connected to the internet using Wi-Fi or Ethernet is used. The gateway then needs to be registered on The Things Network (TTN) to communicate with TTN's cloud infrastructure from where various cloud servers like AWS, Google Cloud and Microsoft Azure can be accessed.

6.4.4 Data analysis and decision support

Data Analytics and AI-based decision-making is used to improve agricultural productivity and enhance resource management. AI models training is broadly classified into two types: Supervised and Unsupervised Learning.

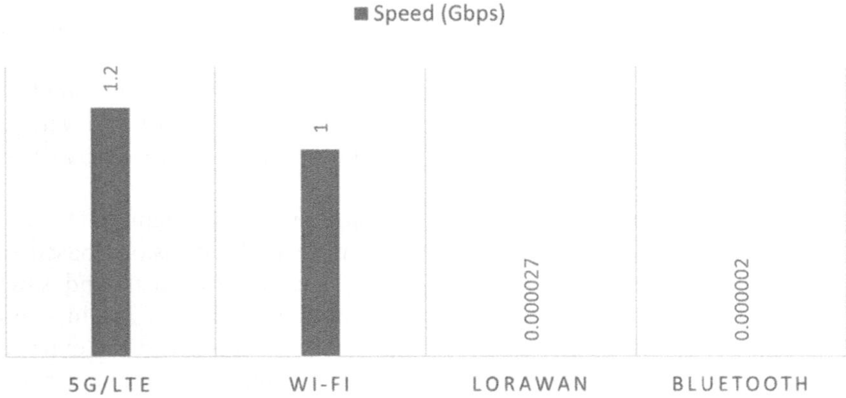

Figure 6.5 Graph representing data transfer speeds of different wireless technologies.

1. Supervised Learning: In supervised learning the training data is labelled. This way the model makes predictions based on features of the input data. For agricultural settings, train the model on the huge datasets that are available online. For example, to detect plant diseases, data labelled with disease name or just labels like "healthy", "mild" and "unhealthy" can be fed to the model. Sometimes, the data might contain outliers and noise that should be cleaned, normalized and standardized. Depending on the nature of output and type of problem classification or regression models can be used. For disease type or stress level measurements of a plant random forest can be implemented which is highly effective for both regression and classification models. It uses multiple decision tress that are trained independently after which their outputs are combined to make more accurate predictions [12].

   Training the Model and Model Evaluation: Split the dataset into training and testing sets. Usually, 80–20 splitting is followed, that is 80% training and 20% testing subsets. After training the model using the labelled data, evaluate the performance using performance metrics like accuracy, F1 score and precision. To improve performance, change the hyperparameters of the model.

   Deployment: Once the model is trained and tested, deploy it on to the edge devices. Choosing the right edge device is very important and is discussed in Section 6.3.3 (Data Processing Layer).
2. Unsupervised Learning: Unsupervised learning technique is used when datasets are not labelled. These use clustering algorithms like *K*-Mean and DBSCAN that group the data with similar features. This can be used for irrigation prediction based on soil moisture levels, humidity levels and temperature [13].

Model Evaluation: Evaluate the results using metrics like Davies-Bouldin Index which measure how well the data fits into the clusters and data separation.

Both supervised and unsupervised learning have unique applications and solve different problems. A hybrid approach that leverages strengths of both the models would be optimal for efficient agricultural practices.

Decision Support System: A decision support system (DDS) is a data-driven tool that helps the farmers to make decisions based on real-time data from various sensors, weather forecasts and satellites. Machine learning models and big data analytics are incorporated to analyse patterns in crop growth with respect to the external factors. The DDS provides an easy-to-use interface for the farmers that includes crop information and action that needs to be taken. Additionally, farmers are notified with alerts for urgent action that needs to be taken. This significantly improves sustainability, reduces risks and increases productivity.

Use Cases and Benefits: Field Mapping and Surveying – Three-dimensional mapping of the field using UAVs increases understanding of crop distribution, crop density and terrain.

Crop Health Monitoring: The IoT sensors and UAV-based monitoring help capture vital plant data that could detect underlying diseases, nutrient deficiencies and pest infestation. This way farmers can spray pesticides in targeted areas rather than covering whole fields, thereby increasing efficient resource utilization and decreasing pollution caused by the chemicals in the pesticides.

Precision Irrigation: While under-irrigation can cause plant dehydration, over-irrigation can wash away nutrients in the soil. It is necessary to supply right amount of water at the right times for optimal yield. This can be achieved using various sensors like soil moisture sensor, temperature sensor and humidity sensors.

AI-Driven Decision-Making: Farmers can make decisions on the basis of AI predictions. Nutrient management and yield prediction trained from satellite imagery and historical data can be used to minimize waste and increase efficiency.

6.5 SUMMARY

Technology plays a crucial role in meeting the food demands of the growing population. Inefficiencies in traditional methods may lead to soil degradation (due to land overuse), water scarcity and pollution (excessive pesticide use). This can be tackled by agricultural innovations like UAV and IoT-based monitoring. These sensors continuously collect data of crops and the environment. These include soil moisture, humidity, multispectral imaging that indicate plant health and nutrition balance. Deploying IoT sensors using UAV-based RSSI scan ensures stable network connectivity. Technologies like cloud computing offer flexibility and remote access to the farmland information, decreasing on-site time and efforts. Edge computing has become a crucial that cleverly offloads processing to the devices near the source to decrease latency and bandwidth usage. AI and machine learning algorithms can forecast outcomes by analysing data from the sensors. This can help farmers make data-driven decisions to optimize crop yield while minimizing required resources. Overall, technology, particularly IoT and UAV-based analysis, increases not only productivity but also sustainability by promoting eco-friendly practices.

REFERENCES

1. Phasinam, Khongdet, et al. "Application of IoT and cloud computing in automation of agriculture irrigation." *Journal of Food Quality* 2022.1 (2022): 8285969.
2. Vazquez-Carmona, E. Viridiana, et al. "Coverage path planning for spraying drones." *Computers & Industrial Engineering* 168 (2022): 108125.
3. Anjum, Mahnoor, et al. "RSSI fingerprinting-based localization using machine learning in LoRa networks." *IEEE Internet of Things Magazine* 3.4 (2021): 53–59.
4. Yu, Limin, et al. "Review of research progress on soil moisture sensor technology." *International Journal of Agricultural and Biological Engineering* 14.4 (2021): 32–42.
5. Velusamy, Parthasarathy, et al. "Unmanned Aerial Vehicles (UAV) in precision agriculture: Applications and challenges." *Energies* 15.1 (2021): 217.
6. Peng, Yao, et al. "Early detection of plant virus infection using multispectral imaging and spatial–spectral machine learning." *Scientific Reports* 12.1 (2022): 3113.
7. Chlingaryan, A., Sukkarieh, S., & Whelan, B. "A review of UAV methods in precision agriculture." *Computers and Electronics in Agriculture* 118 (2015): 1–14.
8. Ikotun, Abiodun M., et al. "*K*-means clustering algorithms: A comprehensive review, variants analysis, and advances in the era of big data." *Information Sciences* 622 (2023): 178–210.
9. Ghazal, Taher M. "Performances of *k*-means clustering algorithm with different distance metrics." *Intelligent Automation & Soft Computing* 30.2 (2021): 735–742.
10. Javornik, Tomislav, et al. "Monitoring drought stress in common bean using chlorophyll fluorescence and multispectral imaging." *Plants* 12.6 (2023): 1386.
11. Gugan, Gopi, and Anwar Haque. "Path planning for autonomous drones: Challenges and future directions." *Drones* 7.3 (2023): 169.

12. Panigrahi, Bharati, Krishna Chaitanya Rao Kathala, and M. Sujatha. "A machine learning-based comparative approach to predict the crop yield using supervised learning with regression models." *Procedia Computer Science* 218 (2023): 2684–2693.
13. Benos, Lefteris, et al. "Machine learning in agriculture: A comprehensive updated review." *Sensors* 21.11 (2021): 3758.

Chapter 7

Cloud-based renewable energy integration in smart agriculture using IoT

Shweta Singh and Rahul Kumar

7.1 INTRODUCTION

Wireless power transfer (WPT) capability is significantly influenced by node positioning, where link distance plays a crucial role in determining energy harvesting (EH) efficiency and outage performance in the sensors of Internet of Things (IoT) systems. This challenge becomes even more pronounced in precision agriculture, where IoT-enabled unmanned aerial vehicles (UAVs) facilitate remote sensing, monitoring, and automated resource distribution. Integrating realistic mobility models for simultaneous wireless information and power transfer (SWIPT) network simulation [1] has been a persistent challenge, particularly in scenarios where UAV movement affects energy efficiency and communication reliability.

In UAV-assisted precision agriculture, mobility dynamics are crucial since EH mechanisms are highly dependent on path loss factors. Specifically, in GHz-range mmWave communications, where smaller cell radii impact connectivity, UAV mobility significantly alters the statistical power distribution of the received signal. Incorporating mobility models into IoT-based UAV networks provides a more accurate analysis of real-world agricultural monitoring applications. However, only a few studies have explored this aspect.

With advancements in IoT-based UAV SWIPT networks improving EH efficiency for far-field transmission, UAV-based node deployment has gained considerable research interest to enhance coverage, accessibility, and communication link reliability [2–4]. In precision agriculture, UAV-assisted wireless networks enable real-time crop health monitoring, livestock tracking, and optimized irrigation. These networks offer extended coverage, increased flexibility, and prolonged network lifespan. However, these advantages come with inherent challenges, particularly due to UAV mobility and drift mechanisms influenced by air turbulence. To ensure accurate performance evaluation, integrating a suitable mobility model is essential for precise statistical analysis of the derived performance metrics.

In this chapter, we extend this analysis considering UAV-assisted SWIPT model for IoT applications in precision agriculture, where UAVs serve as mobile relays for data collection, monitoring, and communication.

DOI: 10.1201/9781003589730-7

Specifically, we consider the mobility of UAVs and ground-based sensor nodes to evaluate the outage performance of the UAV-based IoT system over the fluctuating two-ray (FTR) fading channel. The proposed model is particularly advantageous due to its potential for enhanced energy efficiency, which minimizes the need for manual intervention in IoT-based networks. This characteristic is crucial for agricultural monitoring and control applications, as it facilitates autonomous operation, real-time data acquisition, and optimized resource utilization. By integrating intelligent sensing, edge computing, and adaptive communication protocols, the model supports precision agriculture, enabling automated decision-making for irrigation management, soil health assessment, and crop monitoring while reducing operational costs and human effort. Our analysis is conducted under two scenarios:

1. Mobility in the user (IoT device) node, where the relay-to-destination instantaneous link distance is treated as a random variable.
2. Simultaneous mobility in both the user (IoT device) and UAV-based relay node, where both source-to-relay and relay-to-destination links are considered random. To address the increased complexity in the latter case, we introduce a new parameter, Z, which collectively accounts for mobility-induced variations in link distances.
3. Using above stated approaches, we compared the performance of amplify-and-forward (AF) and decode-and-forward (DF) relays in a cooperative IoT system.

The fundamental mathematical framework for our communication model is inspired by [5–7], but our analysis goes further by comparing their performance considering static and UAV mobility and IoT performance.

The Random Waypoint (RWP) mobility model is widely studied by researchers to analyze the movement patterns of sensor nodes and user equipment in wireless networks [8–12]. The motivation behind choosing the RWP mobility model is because RWP accurately represents the random movement patterns of UAVs in agricultural environments, where autonomous drones monitor fields, track livestock, and collect environmental data. The model accounts for random speed, direction, and pause times, mimicking the unpredictable movements of UAVs navigating farmland. A similar RWP-based mobility model is explored in [13], where a dedicated power beacon enables direct communication between the Hybrid Access Point (H-AP) and the user terminal. However, the mobility dynamics of a cooperative IoT system operating over an FTR fading channel remain unexplored. Our system model builds on the work in [14], which examines UAV-based relay nodes under AF and DF relaying protocols. However, while that study assumes static nodes, our work addresses the critical challenge of mobility—both for UAV-based relay nodes and for ground-based user equipment in agricultural deployments.

In this chapter, we investigate UAV mobility across different topologies—1D, 2D, and 3D—while ensuring operation within a specified link distance range. We further extend our analysis by incorporating maximum ratio combining (MRC) diversity to evaluate the percentage improvement in secrecy outage probability (SOP) under mobile network conditions. This approach provides a more realistic performance evaluation for UAV-assisted IoT applications in precision agriculture, where reliable and energy-efficient communication is essential for large-scale deployment.

The remainder of this chapter is structured as follows. Section 7.2 presents the system model for both the considered cases. Section 7.3 presents the derivation of the SOP parameter for both AF and DF relays considering mobile user terminals for single antenna receivers and MRC multi-antenna receivers. Also, an asymptotic analysis of the derived expression is included. Section 7.4 presents the comparative mathematical framework of SOP computation for AF and DF relays considering a single antenna receiver for the scenario of mobility in both relay and user nodes. Section 7.5 depicts the numerical results and analysis. Finally, the chapter is summarized in Section 7.6.

7.2 UAV-ASSISTED IoT SYSTEM MODEL

Figure 7.1 illustrates a dual-relay communication model, where a H-AP serves as the source, transmitting data to a IoT user (destination) via two different relay strategies: Figure 7.1a: Fixed Relay Node-Based Communication (e.g., a base station or a static repeater tower), Figure 7.1b: UAV-Assisted Mobile Relay Communication. This model is particularly relevant to UAV-enabled IoT networks for precision agriculture, where real-time data acquisition, communication efficiency, and adaptive coverage are crucial for agricultural monitoring and control.

H-AP acts as the central transmitter, which could be an edge server or agricultural IoT gateway responsible for processing and transmitting sensor data. In an agricultural scenario, it collects environmental parameters such as soil moisture, temperature, humidity, and crop health data from distributed IoT sensors. It transmits this data over a wireless medium to the mobile user (farmer or an agricultural processing unit). The mobile aerial relay (UAV) is used in Figure 7.1b for enhanced flexibility and adaptive coverage. In an agricultural scenario, UAVs can dynamically position themselves based on signal strength, terrain conditions, and sensor network topology. The UAV-based relay improves line-of-sight (LoS) connectivity, reducing interference and enhancing throughput. It can optimize energy consumption by adjusting its flight trajectory and hovering at locations where signal strength maximization is required. Mobile User (Destination) represents an end-user or a remote IoT device that receives the transmitted data. The data

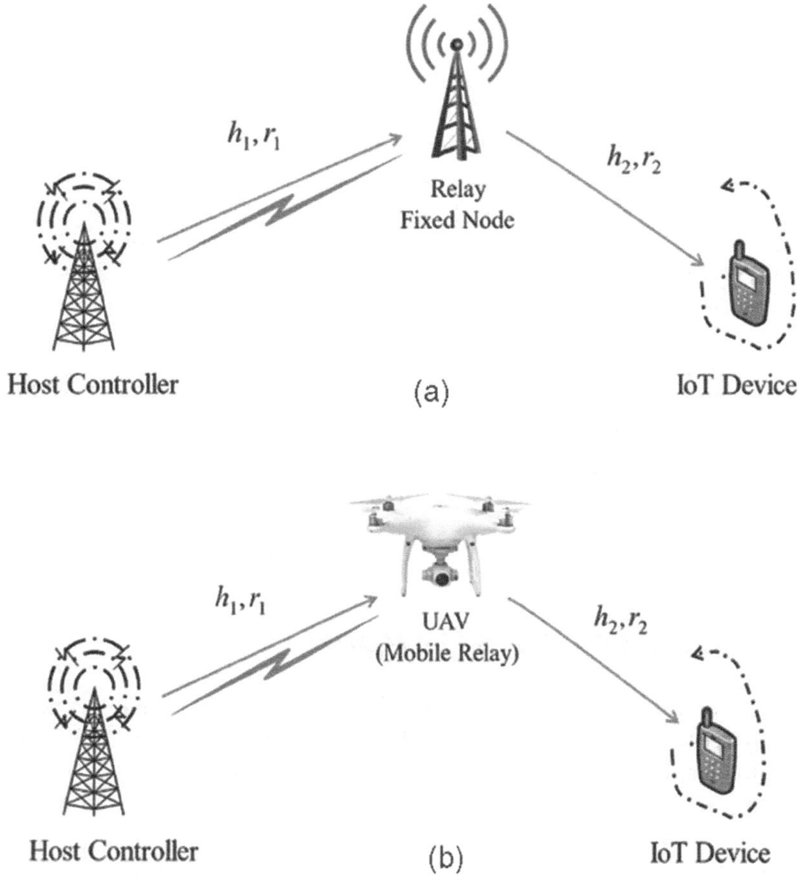

Figure 7.1 Dual-hop SWIPT model with node exhibiting RWP mobility: (a) user node mobile (b) relay and user node mobiles.

is useful for applications like precision irrigation, pest monitoring, and automated fertilization. The relay uses DF or AF techniques to enhance data transmission reliability.

The instantaneous capacity of SWIPT system with Half-Duplex Time-Switching Relaying (HD-TSR) protocol is given as:

$$C_i = \frac{(1-\tau)}{2}\log_2(1+\gamma_i) \tag{7.1}$$

where $i \in (r,d)$ denotes the relay and destination nodes and their respective signal-to-noise ratio (SNR) is given by γ_i. Factor $1/2$ signifies HD relaying operation.

In literature, SOP for both AF and DF relays are broadly studied. SOP is defined as the probability that instantaneous capacity falls below the threshold capacity (C_{th}). The mathematical expression for the SOP for DF and AF relays is given as:

$$SOP_{DF}(C_{th}) = \Pr\left(\min\left\{C_r(\gamma_r), C_d(\gamma_d)\right\} < C_{th}\right) \quad (7.2)$$

$$SOP_{AF}(C_{th}) = \Pr\left(C_d(\gamma_d) < C_{th}\right) \quad (7.3)$$

where γ_r, γ_d represent received SNR at relay and destination nodes and κ, Δ denote the instantaneous capacity at relay and destination nodes, respectively. The linear EH model depicts that the harvested energy is proportional to the received RF signal power. Utilizing the linear EH model, the harvested energy is given as:

$$E_H = \frac{\eta\tau T P_{st}|h_1|^2}{r_1^{\varepsilon}} \quad (7.4)$$

where η is the EH efficiency $(0 < \eta < 1)$ that depends on the relay circuitry, $|\cdot|$ represents mod operator. The received signal at the relay node during the EH phase is expressed as:

$$y_r(t) = \sqrt{\frac{P_{st}}{r_1^{\varepsilon}}} h_1 s(t) + n_r(t) \quad (7.5)$$

where ε depicts the path loss exponent $(2 < \varepsilon < 4)$, P_{st} represents source transmitted power, $s(t)$ is source information signal with $E\left[|s|^2\right] = 1$, and $n_r(t)$ is the narrowband Gaussian noise with the variance σ_r^2. In the next section, we derive a unified mathematical framework for analyzing the SOP performance of this system model. Also, the harvested energy and received signal at the relay node are given by (7.4) and (7.5) respectively. In general, the SOP parameter is evaluated as given in (7.2) and (7.3) respectively, wherein the defined instantaneous capacity is given in (7.1). The mobility of wireless nodes is modeled by utilizing the RWP mobility model derived and illustrated in the subsequent sections for both scenarios.

Case I:

For the presented model in Figure 7.1a, H-AP to relay link is assumed to be static; hence, instantaneous link distance r_1 and maximum link distance R_1 are used interchangeably. The relay to UE or destination link is considered to 7be mobile; hence r_2 represents a random variable following RWP distributions within the maximum coverage of R_2 [13].

RWP random mobility model represented by $\{(P_0,P_1,v_1,\tau_1),(P_1,P_2,v_2,\tau_2)\}$ is popularly used in ad-hoc networks and dynamic network modeling comprising mobile objects. For the RWP mobility model probability density function (PDF) of instantaneous distance r_i is given as [13]

$$f_R(r_i)=\sum_{l_i=1}^{L_i} v_{l_i}\frac{r_i^{\rho_{l_i}}}{R_i^{\rho_{l_i}+1}} \quad 0<r_i\le R_i \tag{7.6}$$

r_i represents the distance between two nodes. A generic dual-hop system i is given by $i\in(1,2)$. Hence, r_1 represents source-to-relay node distance and r_2 represents relay-to-destination node distance. R_i represents the radius of maximum coverage distance by the respective nodes, where the maximum radial distance from the source to relay is given by R_1 and the maximum radial distance from the relay to destination is given by R_2. ρ_{l_i} and v_i are RWP coefficient parameters derived in [15] for 1D, 2D, and 3D network topologies corresponding to RWP and RD mobility model. In (7.6) the coefficients like ρ_l, v_l, l_i for 1D, 2D, and 3D mobility topologies are obtained. Thus the equation given in (7.6) can be precisely rewritten as:

$$f_R(r_2)=\sum_{l^i=1}^{L} v_l\frac{r_2^{\rho_l}}{R_2^{\rho_l+1}} \quad 0\le r_2\le R_2 \tag{7.7}$$

Case II:

Due to the mobility of the relay and destination nodes, the H-AP to relay and relay to receiver links are regarded as randomly fluctuating for the model shown in Figure 7.1b; hence both r_1 and r_2 link depicts the random variable with RWP distribution given in (7.6). On normalizing r_i with respect to maximum coverage distance, i.e., $r_{n_i}=r_i/R_i$ and transforming PDF in (7.6), we get

$$f_R(r_{n_i})=\sum_{l_i=1}^{L_i} v_{l_i} r_{n_i}^{\rho_{l_i}} \quad 0<r_{n_i}\le 1 \tag{7.8}$$

In the presented analysis, it is assumed that both the channels experience independent FTR fading. The SNR (γ_i) expression for the FTR fading channel PDF and cumulative distribution function (CDF) is given in (7.13) and (7.14) respectively. For ease of computation here, we assume $1/2\sigma_i^2=\mu_i$ that indicates the overall power linked to the scattered components. In the next section, we derive a unified mathematical framework for analyzing the SOP performance of this presented system model under two case studies.

7.3 DERIVATION OF SYSTEM OUTAGE PROBABILITY CONSIDERING MOBILE USER NODE

This section presents the SOP derivation for the given communication setup considering randomly moving user terminal. The mathematical SOP expression for DF and AF-based relaying as given in (7.2) and (7.3) for the static link scenario is re-formulated in (7.9) for the mobile link scenario. The average received power or average received SNR, however, is no longer constant when taking into account mobile relay-to-destination link; rather, it is dependent on the nodal distance r_2. Thus, conditional SOP can be described as follows:

$$SOP_{DF}(C_{th}) = \Pr\left(\min\{C_r(\gamma_r), C_d(\gamma_d)\} < C_{th} | r_2\right) \tag{7.9}$$

$$SOP_{AF}(C_{th}) = \Pr\left(C_d(\gamma_d) < C_{th} | r_2\right) \tag{7.10}$$

Given the HD-TSR protocol, (7.1) provides the instantaneous capacity, i.e., used to find SOP from (7.9) and (7.10) for DF and AF relays. For generality here, the equation for received power can be used in the same way as received SNR since we assume unit noise power; in other words, received power and received SNR can be used interchangeably. Assuming the relay-to-destination link to be mobile the r_2 distance is assumed to vary randomly and hence SOP is evaluated by averaging over random variable r_2

$$SOP_{DF}^{dyn}(C_{th}) = \int_0^{R_2} \Pr\left(\min\{C_r(\gamma_r), C_d(\gamma_d)\} < C_{th} | r_2\right) f_R(r_2)\, dr_2 \tag{7.11}$$

$$SOP_{AF}^{dyn}(C_{th}) = \int_0^{R_2} \Pr\left(C_d(\gamma_d) < C_{th} | r_2\right) f_R(r_2)\, dr_2 \tag{7.12}$$

7.3.1 DF relaying with single antenna receiver

In this section, SOP expression for the SWIPT system model is derived for DF-TSR protocols while considering the random mobility of the user node equipped with a single antenna. The derived result provides significant insight into the presented system under study and provides a baseline for the improved receiver structure with a diversity scheme. The FTR fading channel PDF and CDF expressions are given in (7.13) and (7.14). The PDF and CDF of instantaneous received power for N_T terms can be rewritten as [16]

$$f_{\|h_i\|^2}(\gamma) = \sum_{j_i=0}^{N_T} d_{j_i} \frac{\gamma^{j_i} e^{-\gamma/2\sigma_i^2}}{\Gamma(j_i+1)\left(2\sigma_i^2\right)^{j_i+1}} \tag{7.13}$$

$$F_{\|h_i\|^2}(\gamma) = \sum_{j_i=0}^{N_T} d_{j_i} \frac{\Upsilon\left(j_i + 1, \gamma / 2\sigma_i^2\right)}{\Gamma\left(j_i + 1\right)} \tag{7.14}$$

Further, the analysis incorporates the random mobility in the user node with a random variable r_2 link whose PDF is given by RWP mobility as in (7.7). Here, the system deploys a dual-hop link so the time allocation for WIT phase from source to relay and relay to the destination node is $(1-\tau)T/2$. In the case of DF relaying the received signal at the destination node is represented as follows:

$$y_d(t) = \sqrt{\frac{P_{rt}}{r_2^{\varepsilon}}} h_2 \bar{s}(t) + n_d(t) \tag{7.15}$$

where $\bar{s}(t)$ is the decoded source signal and $n_d(t) = n_r(t) + n_c(t)$ represents overall noise at the destination node with the variance of σ_d^2, $n_c(t)$ is the noise by the information decoder due to RF to baseband conversion. Transmitted power from relay node (P_{rt}) during the information transmission phase is computed utilizing (7.4) as follows:

$$P_{rt} = \frac{E_H}{(1-\tau)T/2} = \frac{2\eta\tau P_{st} |h_1|^2}{(1-\tau) r_1^{\varepsilon}} \tag{7.16}$$

From (7.15) and (7.16) we have:

$$y_d(t) = \sqrt{\frac{2\eta\tau P_{st}}{(1-\tau) r_1^{\varepsilon} r_2^{\varepsilon}}} h_1 h_2 \bar{s}^{i}(t) + n_d(t) \tag{7.17}$$

Using (7.5) and (7.17) we find the received instantaneous SNR at the relay (γ_r) and destination node (γ_d) as (7.18) and (7.19) respectively. Here for generality, we assume $\sigma_r^2 = \sigma_d^2 = 1$.

$$\gamma_r = \frac{P_{st} |h_1|^2}{r_1^{\varepsilon} \sigma_r^2} \tag{7.18}$$

$$\gamma_d(t) = \frac{2\eta\tau P_{st} |h_1|^2 |h_2|^2}{(1-\tau) r_1^{\varepsilon} r_2^{\varepsilon} \sigma_d^2} \tag{7.19}$$

We utilize the expression for the received SNR at the relay and destination node that will be utilized for evaluating SOP subsequently. Now rewrite (7.11) by mathematical reformulation,

$$SOP_{\mathrm{DF}}^{dyn}(C_{th}) = \int_{0}^{R_2} \left[1 - \Pr\left(\min\{C_r(\gamma_r), C_d(\gamma_d)\} \geq C_{th} | r_2\right)\right] f_R(r_2)\, dr_2 \tag{7.20}$$

$$SOP_{\mathrm{DF}}^{dyn}(C_{th}) = \int_{0}^{R_2} \begin{bmatrix} 1 - \Pr\left(C_r(\gamma_r) \geq C_{th} | r_2\right) \\ + \Pr\left(C_r(\gamma_r) \geq C_{th}, C_d(\gamma_d) < C_{th} | r_2\right) \end{bmatrix} f_R(r_2)\, dr_2$$

Now using area property for PDF $f_R(r_2)$ and the fact that probability $\Pr\left(C_r(\gamma_r) \geq C_{th} | r_2\right)$ is independent of parameter r_2 (7.20) is rewritten as

$$SOP_{\mathrm{DF}}^{dyn}(C_{th}) = \quad 1 - \underbrace{\Pr\left(C_r(\gamma_r) \geq C_{th}\right)}_{P_1(C_{th})} + \int_{0}^{R_2} \underbrace{\Pr\left(C_r(\gamma_r) \geq C_{th}, C_d(\gamma_d) < C_{th} | r_2\right) f_R(r_2)\, dr_2}_{P_2(C_{th})} \tag{7.21}$$

The PDF of the random variable r_2 (7.21) is incorporated as the statistical average over the entire maximal radial range. From (7.21), to compute $SOP_{\mathrm{DF}}^{dyn}(C_{th})$ it is essential to evaluate probabilities $P_1(C_{th})$ and $P_2(C_{th})$. To evaluate SOP using (7.2) and using some mathematical reformulation:

$$SOP_{\mathrm{DF}}^{stat}(C_{th}) = 1 - \Pr\left(\min\{C_r(\gamma_r), C_d(\gamma_d)\} \geq C_{th}\right) \tag{7.22}$$

Now (7.22) can be re-represented as

$$SOP_{\mathrm{DF}}^{stat}(C_{th}) = 1 - \underbrace{\Pr\left(C_r(\gamma_r) \geq C_{th}\right)}_{P_1(C_{th})} + \underbrace{\Pr\left(C_r(\lambda_r) \geq C_{th}, C_d(\gamma_d) < C_{th}\right)}_{P_2(C_{th})} \tag{7.23}$$

It is deduced from (7.23) that $SOP_{\mathrm{DF}}^{stat}(C_{th})$ require computation of two probabilities $P_1(C_{th})$ and $P_2(C_{th})$, where $P_1(C_{th})$ denotes H-AP to relay link outage probability and $P_2(C_{th})$ denotes relay-to-destination link outage probability. Now substituting $X = h_1^2$ into (7.18) and using (7.1), $P_1(C_{th})$ can be rewritten as:

$$\begin{aligned} P_1(C_{th}) &= \Pr\left(\frac{(1-\tau)}{2}\log_2\left(1 + \frac{P_{st}X}{r_1^{\varepsilon}\sigma_r^2}\right) \geq C_{th}\right) = \Pr\left(\frac{P_{st}X}{r_1^{\varepsilon}\sigma_r^2} \geq \upsilon\right) \\ &= 1 - F_X(a_1\upsilon) \end{aligned} \tag{7.24}$$

where $\upsilon = 2^{\frac{2C_{th}}{1-\tau}} - 1$, $a_1 = \frac{r_1^{\varepsilon}\sigma_r^2}{P_{st}}$, which directly signifies that increasing link distance and path loss and decreasing transmit power outage will increase.

$F_X(\cdot)$ is the CDF of the random variable X. Since X is FTR distributed, its CDF is given by (7.14). Hence substituting (7.14) into (7.24) $P_1(C_{th})$ is obtained as

$$P_1(C_{th}) = 1 - \sum_{j_1=0}^{N_T} \frac{d_{j_1}}{\Gamma(j_1+1)} \Upsilon\left(j_1+1, a_1 v / 2\sigma_1^2\right) \tag{7.25}$$

Now substituting $X = h_1^2$ and $Y = h_2^2$ into (7.19) and using (7.1), $P_2(C_{th})$ is expressed as

$$P_2(C_{th}) = \Pr\left(X \geq a_1 v, \frac{2\eta\tau P_{st} XY}{(1-\tau) r_1^{\varepsilon} r_2^{\varepsilon} \sigma_d^2} < v\right) = \int_{a_1 v}^{\infty} f_X(z) F_Y\left(\frac{a_2 v}{z}\right) dz \tag{7.26}$$

where $a_2 = \frac{(1-\tau) r_1^{\varepsilon} r_2^{\varepsilon} \sigma_d^2}{2\eta\tau P_{st}}$, now substituting PDF and CDF of FTR distributed random variables X and Y from (7.13) and (7.14) respectively we obtain

$$P_2(C_{th}) = \sum_{j_1=0}^{N_T} \sum_{j_2=0}^{N_T} \frac{d_{j_1} d_{j_2}}{\Gamma(j_1+1)\Gamma(j_2+1)} \frac{1}{\left(2\sigma_1^2\right)^{j_1+1}} \tag{7.27}$$

$$\times \underbrace{\int_{a_1 v}^{\infty} z^{j_1} e^{-z/2\sigma_1^2} \Upsilon\left(j_2+1, \frac{a_2 v}{2\sigma_2^2 z}\right) dz}_{I_1}$$

A closed-form expression $P_2(C_{th})$ can be obtained by solving the integral I_1 in (7.27) as given below.

$$P_2(C_{th}) = \sum_{j_1=0}^{N_{T_1}} \sum_{j_2=0}^{N_{T_2}} \frac{d_{j_1}}{\Gamma(j_1+1)} \frac{d_{j_2}}{\Gamma(j_2+1)} \frac{1}{\left(2\sigma_1^2\right)^{j_1+1}}$$

$$\times \left\{ \begin{array}{l} \dfrac{\Gamma(j_1+1)\Gamma(j_1+1, a_1 v)}{\mu_1^{j_1+1}} - \Gamma(j_2+1)(a_1 v)^{j_1+1} \\ \displaystyle\sum_{n=0}^{[j_2]} \frac{1}{n!} \left(\frac{\mu_2 a_2}{a_1}\right)^n G_{1,1;1,0;0,1}^{1,0;0,1;1,0}\left(\begin{array}{c} n-j_1 \\ n-j_1+1 \end{array} \middle| \begin{array}{c} 1 \\ - \end{array} \middle| \begin{array}{c} - \\ 0 \end{array} \middle| \frac{1}{a_1 v}, \frac{\mu_2 a_2}{a_1} \right) \end{array} \right\} \tag{7.28}$$

Proof: *See Appendix A*

where symbol $\lceil x \rceil$ denotes the nearest integer greater than or equal to x, and function $G_{\cdot,\cdot;\cdot,\cdot;\cdot,\cdot}^{\cdot,\cdot;\cdot,\cdot;\cdot,\cdot}\left(\cdot|\cdot|\cdot|\cdot,\cdot\right)$ represents extended generalized bivariate Meijer-G function (EGBMGF) [17]. EGBMGF is a generalized function that can express the majority of the known functions as its special case by changing its variable and parameter values. In (7.28) EGBMGF is applied as in this case the integral ranges from constant value to infinity which leads to a bivariate Meijer-G form. Now, utilizing (7.25) and (7.28) in (7.23) one can obtain SOP_{DF}^{stat} expression to evaluate the performance of the dual-hop DF-TSR-based SWIPT system under a static node scenario. Applying similar operations as in (7.24) and (7.25) we compute $P_1(C_{th})$ which remains the same as in the case of the static scenario. Next, as r_2 is a random variable term due to mobility in the user node, its PDF is given by RWP distribution. Hence, to compute $P_2(C_{th})$ we introduce an additional PDF term $f_R(r_2)$ and integrate it over the entire radial coverage range R_2. To compute $P_2(C_{th})$ substituting $X = |h_1|^2$ and $Y = |h_2|^2$ into (7.18) and (7.19) and using relation for HD-TSR instantaneous capacity relation using (7.1), $P_2(C_{th})$ is given as:

$$P_2(C_{th}) = \int_0^{R_2} \Pr\left(X \geq a_1 v, \frac{2\eta\tau P_{st} XY}{(1-\tau) r_1^{\varepsilon} r_2^{\varepsilon} \sigma_d^2} < v \middle| r_2 \right) f_R(r_2) dr_2 \tag{7.29}$$

$$= \int_0^{R_2} \int_{a_1 v}^{\infty} f_X(z) F_Y\left(\frac{a_2 v r_2^{\varepsilon}}{z} \right) f_R(r_2) dz dr_2$$

where $a_2 = \frac{(1-\tau) r_1^{\varepsilon} \sigma_d^2}{2\eta\tau P_{st}}$; substituting PDF, CDF for X and Y from (7.13) and (7.14) we obtain:

$$P_2(C_{th}) = \sum_{j_1, j_2 = 0}^{N_T} \sum_{l=1}^{L} \frac{v_l}{R_2^{\rho l+1}} \frac{d_{j_1} d_{j_2}}{\Gamma(j_1+1)\Gamma(j_2+1)} \frac{1}{(2\sigma_1^2)^{j_1+1}} \tag{7.30}$$

$$\underbrace{\int_0^{R_2} r_2^{\rho l} \int_{a_1 v}^{\infty} z^{j_1} e^{-z/2\sigma_1^2} \Gamma\left(j_2+1, \frac{a_2 v r_2^{\varepsilon}}{2\sigma_2^2 z} \right) dz dr_2}_{I_7}$$

A closed-form expression of $P_2(C_{th})$ can be obtained by solving the integral I_7 in (7.30) given as:

$$P_2(C_{th}) = \sum_{j_1, j_2 = 0}^{N_T} \sum_{l=1}^{L} \frac{v_l d_{j_1} d_{j_2}}{\Gamma(j_1+1)\Gamma(j_2+1)} \frac{1}{\varepsilon w (2\sigma_1^2)^{j_1+1}} \tag{7.31}$$

$$\times \left(f\left(j_1, \left[j_2+1 \right]\right) - \left(a_2 v R_2^{\varepsilon} / 2\sigma_2^2 \right)^{-w} f\left(j_1 + w, \left[j_2 + w + 1 \right] \right) \right)$$

In (7.28), it can be seen that, with the increase in $a_1 v$, the Meijer-G function decreases non-linearly, leading to a high outage probability $P_2(C_{th})$; thus, SOP increases. Also, due to the presence of a_2 term in polynomial form, the product of both link distances plays a vital role (since $a_2 = (1-\tau) r_1^{\varepsilon} r_2^{\varepsilon} \sigma_d^2 / 2\eta\tau P_{st}$). This infers that a_2 will be maximum when both the link distances are the same, i.e., $r_1 = r_2$ and minimum when either of the link distances is low, i.e., $r_1 < r_2$ or $r_2 < r_1$. The former case leads to the situation when the relay is equidistant from both H-AP and the destination node, and then $P_2(C_{th})$ link probability is maximum; thus, SOP increases. For the latter case, the relay is either close to H-AP or the destination node, the $P_2(C_{th})$ link probability reduces; thus, SOP improves. The impact of the parameters is presented in Table 7.1 for better clarity.

Proof: *See Appendix B*

Equation (7.31) denotes that for a high value of ε, $P_2(C_{th})$ increases and hence SOP deteriorates. Here, $w = (\rho_l + 1)/\varepsilon$ and function $f(j_1, \lceil j_2 + 1 \rceil)$ is defined as:

$$f(j_1, \lceil j_2 + 1 \rceil) = (2\sigma_1^2)^{j_1+1} \Gamma(j_1 + 1, a, v) \tag{7.32}$$

$$-(a_1 v)^{j_1+1} \Gamma(j_2 + 1) \sum_{n=0}^{\lceil j_2 \rceil} \frac{1}{n!} \left(\frac{a_2 R_2^{\varepsilon}}{a_1 2\sigma_2^2} \right)^n$$

$$G_{1,1;1,0;0,1}^{1,0;0,1;1,0} \left(\begin{array}{c|c|c|c} n - j_1 & 1 & - & \frac{2\sigma_1^2}{a_1 v}, \frac{a_2 R}{a_1 2c} \\ n - j_1 + 1 & - & 0 & \end{array} \right.$$

Denotation $\lceil x \rceil$ represents the nearest integer, i.e., equal to or larger than x, and function $G_{\cdot,\cdot;\cdot,\cdot;\cdot,\cdot}^{\cdot,\cdot;\cdot,\cdot;\cdot,\cdot}(\cdot|\cdot|\cdot|\cdot,\cdot)$ represents EGBMGF [17]. Now, $SOP_{DF}^{dyn}(C_{th})$ for the DF relaying can be obtained by substituting (7.25), (7.31), and

Table 7.1 Impact of different system parameters on

System parameter	*$P_2(C_{th})$ first link outage probability*
σ_d^2 noise variance	For high σ_d^2, $P_2(C_{th})$ would be high
$r_1 r_2$ multiplicative factor of link distances	For maximum value of product of $r_1 r_2$, i.e., when $r_1 = r_2$, $P_2(C_{th})$ would be high
ε path loss	For high ε, $P_2(C_{th})$ would be high
C_{th} threshold capacity	For high C_{th}, $P_2(C_{th})$ would be high
η EH efficiency	For high η, $P_2(C_{th})$ would be low
P_{st} source transmit power	For high P_{st}, $P_2(C_{th})$ would be low

(7.32) into (7.21). Obtained expression for SOP_{DF}^{dyn} can be used to evaluate the performance of the dual-hop SWIPT-based DF relaying protocol for TSR EH.

7.3.2 AF relaying with single antenna receiver

This section derives SOP expression for the SWIPT system model considering AF-TSR protocols with the random mobility of user node, i.e., equipped with a single antenna. The harvested energy from H-AP and the received signal at the relay node are given as in (7.4) and (7.5). The transmitted signal through the AF relay $r_t(t)$ and the received signal at the destination node. Further using (7.19) for the AF relay, the received SNR at the destination node γ_d is given as

$$\gamma_d = \frac{2\eta\tau P_{st}|h_1|^2|h_2|^2}{2\eta\tau r_1^{\varepsilon}\sigma_r^2|h_2|^2 + (1-\tau)r_1^{\varepsilon}r_2^{\varepsilon}\sigma_d^2} = \frac{b_1XY}{b_3Y + b_2r_2^{\varepsilon}} \tag{7.33}$$

where $b_1 = 2\eta\tau P_{st}$, $b_2 = (1-\tau)r_1^{\varepsilon}\sigma_d^2$, $b_3 = 2\eta\tau r_1^{\varepsilon}\sigma_r^2$, $X = |h_1|^2$, and $Y = |h_2|^2$. Now using (7.1), (7.19), and (7.12) SOP for the AF relaying for the dynamic network is evaluated as follows:

$$\begin{aligned} SOP_{AF}^{dyn}(C_{th}) &= \int_0^{R_2} \Pr\left(\frac{b_1XY}{b_3Y + b_2r_2^{\varepsilon}} < v \middle| r_2\right) f_R(r_2)\,dr_2 \\ &= \int_0^{R_2}\int_0^{\infty} F_X\left(\frac{vb_3}{b_1} + \frac{b_2vr_2^{\varepsilon}}{b_1z}\right) f_Y(z) f_R(r_2)\,dz\,dr_2 \end{aligned} \tag{7.34}$$

Hence, substituting (7.13), (7.14) into (7.34) SOP_{AF}^{dyn} is evaluated as

$$\begin{aligned} SOP_{AF}^{dyn} = & \sum_{j_{1,2}=0}^{N_T}\sum_{l=1}^{L} \frac{v_l}{R_2^{\rho_l+1}} \frac{d_{j_1}d_{j_2}}{\Gamma(j_1+1)\Gamma(j_2+1)} \frac{1}{\left(2\sigma_2^2\right)^{j_2+1}} \\ & \times \underbrace{\int_0^{R_2} r_2^{\rho_l} \int_0^{\infty} z^{j_2} e^{-z/2\sigma_2^2}\Upsilon\left(j_1+1, \left(\frac{vb_3}{b_1 2\sigma_1^2} + \frac{b_2vr_2^{\varepsilon}}{b_1 2\sigma_1^2 z}\right)\right) dz\,dr_2}_{I_8} \end{aligned} \tag{7.35}$$

A closed-form expression for integral I_8 is given in Appendix C. Using (7.35) and I_8 expression obtained in Appendix C and PDF area properties, i.e., $\sum_{l=1}^{L}\frac{v_l}{\rho_l+1} = 1$ and $\sum_{j=0}^{N_T} d_j = 1$ a closed-form expression for SOP_{AF}^{dyn} is given as

$$SOP_{AF}^{dyn} = 1 - e^{-\left(\frac{vb_3}{b_1 2\sigma_1^2}\right)} \sum_{j_{1,2}=0}^{N_T} \sum_{l=1}^{L} \frac{v_l}{R_2^{\rho_l+1}} \frac{d_{j_1} d_{j_2}}{\Gamma(j_2+1)} \sum_{n=0}^{j_1} \sum_{m=0}^{n} \binom{n}{m} \left(\frac{vb_3}{b_1 2\sigma_1^2}\right)^{n-c_l}$$

$$\times \left(\frac{b_2}{b_3}\right)^{m-c_l} \frac{\left(2\sigma_2^2\right)^{\omega+c_l-j_2-1}}{n!\varepsilon} G_{1,3}^{2,1}\left(\frac{b_2 v R_2^{\varepsilon}}{b_1 2\sigma_1^2 2\sigma_2^2} \middle| \begin{matrix} 2 \\ c_l, \omega + c_l, 0 \end{matrix} \right) \tag{7.36}$$

where ω is the nearest integer to fraction $(j_2 - n + 1)$ and $c_l = (\rho_l + m\varepsilon + 1)/\varepsilon$. Equation (7.36) provides the closed-form formulation of SOP_{AF}^{dyn} for assessing the effectiveness of the dual-hop AF relaying technique based on SWIPT for TSR EH. The derived expression in (7.36) provides the reference study to further extend our analysis by introducing MRC diversity for multiple antenna reception for AF relay under a dynamic network that is subsequently analyzed in Section 7.3.4.

Due to the exponential term e^{-vb_3/b_1} in (7.36), its associated parameters predominantly affect the SOP performance as given that $b_1 \propto P_{st}$, $b_3 \propto r_1^{\varepsilon}$, $b_2 \propto r_1^{\varepsilon} r_2^{\varepsilon}$, and $v = 2^{2C_{th}/1-\tau} - 1$. This makes the second term in (7.36) an increasing function of P_{st}, thus for high P_{st}, SOP would be low, and a decreasing function of r_1^{ε}, and C_{th}, thus for high r_1^{ε}, and C_{th}, SOP would be high. The r_1^{ε} is the term is predominantly present in SOP expression, denoting source-to-relay link distance has more influence on SOP performance. It signifies that the placement of the relay node closer to the H-AP will yield superior outage performance compared to its counterpart due to better EH capability. The detailed impact of these parameter variations is given in Table 7.2.

7.3.3 DF relaying with MRC multiple antenna receiver

This section extends the analysis presented in Section 7.3.1 (for DF-TSR) to multiple antenna diversity cases. In this section, we introduce the MRC

Table 7.2 Impact of different system parameters on SOP performance of AF relay

System parameter	*SOP performance of AF-TSR based SWIPT system*
σ_r^2, σ_d^2 noise variance	For high σ_r^2, σ_d^2, SOP would be large
$r_1 r_2$ multiplicative factor of link distances	For the maximum value of the product of $r_1 r_2$, i.e., when $r_1 = r_2$, SOP would be large
ε path loss	For high ε, SOP would be large
C_{th} threshold capacity	For high C_{th}, SOP would be large
η EH efficiency	For high η, SOP would be low
P_{st} source transmit power	For high P_{st}, SOP would be low

diversity technique over the FTR fading channel for a mobile network scenario considering the randomly moving mobile user. We have considered *i.i.d* FTR random variable whose PDF and CDF expressions are for the instantaneous SNR at the output of H branch MRC receiver. The transmission scenario was analyzed under different mobility topologies (1D, 2D, and 3D) considering the RWP mobility model. A generalized SOP expression is derived, to evaluate the SWIPT system outage performance in the presented scenario.

The received signal at the destination with H multiple antennas and also with MRC diversity is obtained combining the received SNR at the combiner output [6]. By applying similar steps from [6], $P_{1\_mul}(C_{th})$ can be obtained. Observe as $P_{1\_mul}(C_{th})$ depends on source-to-relay link, which remains the same as in static case, hence given in a similar manner as discussed in [6], such that

$$P_{1\_mul}(C_{th}) = 1 - \sum_{j_1=0}^{N_T} \frac{d_{j_1}}{(j_1+1)} \Upsilon\left(j_1+1, a_1 v / 2\sigma_1^2\right)$$

Further, $P_{2\_mul}(C_{th})$ is computed by substituting $X = |h_1|^2$, $Y = \|h_2\|^2$. Using similar steps as obtained in (7.30), we get

$$P_{2\_mul}(C_{th}) = \sum_{j=0}^{N_T} \sum_{j_2,..j_H=0}^{N_T} \sum_{l=1}^{L} \frac{v_l}{R_2^{\rho_l+1}} \frac{d_j \prod_{p=1}^{H} d_{j_p}}{\Gamma(j+1)\Gamma(J+H)} \frac{1}{\left(2\sigma_1^2\right)^{j+1}} \times \int_0^{R_2} r_2^{\rho_l} \int_{a_1 v}^{\infty} z^j e^{-z/2\sigma_1^2} \Upsilon\left(J+H, \frac{a_2 v r_2^{\varepsilon}}{2\sigma_2^2 z}\right) dz dr_2 \tag{7.37}$$

Integral in (7.37) can be solved using Appendix B and with some mathematical manipulations, we get

$$P_{2\_mul}(C_{th}) = \sum_{j=0}^{N_T} \sum_{j_1,j_2,..j_H}^{N_T} \sum_{l=1}^{L} \frac{v_l d_j \prod_{p=1}^{H} d_{j_p}}{\Gamma(j+1)\Gamma(J+H)} \frac{1}{\left(2\sigma_1^2\right)^{j+1}} \times \frac{1}{\varepsilon w} \left(f\left(j, \left[J+H\right]\right) - \left(\frac{a_2 v R_2^{\varepsilon}}{2\sigma_2^2}\right)^{-w} f\left(j+w\left[J+H+w\right]\right) \right) \tag{7.38}$$

where $w = (\rho_l + 1)/\varepsilon$ and function $f\left(j, \left[J+H\right]\right)$ is defined as

$$f\left(j,\left\lceil J+H\right\rceil\right)=\Gamma\left(J+H\right)\Gamma\left(j+1,a,v\right)\left(2\sigma_1^2\right)^{j+1}-\Gamma\left(J+H\right)\left(a,v\right)^{j+1} \tag{7.39}$$

$$\sum_{n=0}^{\lceil J+H-1\rceil}\frac{1}{n!}\left(\frac{a_2 R_2^{\varepsilon}}{2\sigma_2^2 a_1}\right)^n G_{1,1,1;,0;0,1}^{1,0;0,1;1,0}\left(\begin{array}{c|c|c|c} n-j & 1 & - & \frac{2\sigma_1^2}{a_1 v},\frac{a_2 R_2^{\varepsilon}}{2\sigma_2^2 a_1} \\ n-j+1 & - & 0 & \end{array}\right)$$

It is seen that with the high number of diversity branches, $P_{2\_mul}\left(C_{th}\right)$ reduces, thus outage performance improves.

7.3.4 AF relaying with MRC multiple antenna receiver

This section extends the analysis presented in Section 7.3.2 (for AF-TSR) to multiple antenna diversity case. The harvested energy and the received signal at the relay node are given as in (7.4) and (7.5). Also, the received signal at the destination node with H multiple antennas are represented as $\bar{y}_d(t)$ given in [6]. Using (7.16), substitute P_{rt} into the received SNR at the destination node represented as

$$\gamma_{d\_mrc}^{dyn\_AF}=\frac{2\eta\tau P_{st}\left|h_1\right|^2\left\|h_2\right\|^2}{2\eta\tau r_1^{\varepsilon}\sigma_r^2\left\|h_2\right\|^2+\left(1-\tau\right)r_1^{\varepsilon}r_2^{\varepsilon}\sigma_d^2}=\frac{b_1 XY}{b_3 Y+b_2 r_2^{\varepsilon}}, \tag{7.40}$$

where $b_1=2\eta\tau P_{st}$, $b_2=\left(1-\tau\right)r_1^{\varepsilon}\sigma_d^2$, $b_3=2\eta\tau r_1^{\varepsilon}\sigma_r^2$, $X=\left|h_1\right|^2$, and $Y=\left\|h_2\right\|^2$. Now applying similar steps as to get (7.35), $SOP_{AF\_mrc}^{dyn}$ is given as:

$$SOP_{AF\_mrn}^{dyn}=\sum_{j=0}^{N_T}\sum_{j_1,j_2,..j_H=0}^{N_T}\sum_{l=1}^{L}\frac{v_l}{R_2^{\rho_l+1}}\frac{d_j\prod_{p=1}^{H}d_{j_p}}{\Gamma\left(j+1\right)\Gamma\left(J+H\right)}\frac{1}{\left(2\sigma_2^2\right)^{J+H}} \tag{7.41}$$

$$\int_0^{R_2} r_2^{\rho_l}\int_0^{\infty} z^{J+H-1}e^{-z/2\sigma_2^2}\Upsilon\left(j+1,\frac{1}{2\sigma_1^2}\left(\frac{vb_3}{b_1}+\frac{b_2 v r_2^{\varepsilon}}{b_1 z}\right)\right)$$

with the aid of Appendix C and with area properties of PDF, i.e., $\sum_{l=1}^{L}\frac{v_l}{\rho_l+1}=1$, $\sum_{j_1,j_2,...j_H=0}^{N_T}\prod_{p=1}^{H}d_{j_p}=1$, and $\sum_{j=0}^{N_T}d_j=1$ a closed-form expression for $SOP_{AF\_mrc}^{dyn}$ is given as

$$I_7 = \frac{R_2^{\rho+1}}{\varepsilon\omega}\underbrace{\int_0^{\infty} z^{j_1} e^{-\alpha z}\Upsilon\left(j_2+1, \frac{\beta R_2^{\varepsilon}}{z}\right)dz}_{I_9} - \left(\beta R_2^{\varepsilon}\right) \tag{7.42}$$

$$-\omega \frac{R_2^{\rho+1}}{\varepsilon\omega}\underbrace{\int_0^{\infty} z^{j_1} e^{-\alpha z}\Upsilon\left(\omega + j_2+1, \frac{\beta R_2^{\varepsilon}}{z}\right)dz}_{I_{10}}.$$

where $c_l = \left(\rho_l + m\varepsilon + 1\right)/\varepsilon$ and w is the nearest integer to fraction $J + H - n$. The inference drawn from (7.42) suggests that as H increases, the overall $SOP_{AF\_mrc}^{dyn}$ reduces.

7.4 SYSTEM OUTAGE PROBABILITY CONSIDERING MOBILE USER AND RELAY NODE

In this section, a more complicated scenario of case II is analyzed that considers the influence of simultaneous mobility in both the user and relay nodes. To envisage the simultaneous mobility impact of both wireless nodes causing the random link variation for both links the mathematical problem is formulated. In order to tackle high-end mathematical complexity and to avoid multiple integrations in this case, we introduce a random variable Z, defined as the ratio of random variable $Z_1 = \gamma_i / \chi_i$, $Z_i > 0$, where γ_i is the received instantaneous SNR over the FTR fading channel with PDF $f_\gamma\left(\gamma_i\right)$ and $\chi_i = r_{n_i}^{\varepsilon_i}$ with r_{n_i} representing the normalized link distance with PDF $f_R\left(r_{n_i}\right)$ given as in (7.8).

The significance of parameter Z lies in the fact that it simultaneously analyzes the impact of node mobility on (i) received instantaneous SNR γ_i, (ii) link distance r_i, and (iii) path loss exponent ε_i on system outage performance parameter. The introduced parameter has extensive application in system performance analysis where node mobility is predominant.

7.4.1 CDF $F_Z\left(z_i\right)$ for the random variable Z_i

The PDF of the random variable Z is given as:

$$F_Z\left(z_i\right) = \Pr\left(Z_i < z_i\right) = \Pr\left(\frac{\gamma_i}{\chi_i} < z_i\right) = \Pr\left(\gamma_i < z_i\chi_i\right) = \int_0^1 F_\gamma\left(z_i\chi_i\right) f_\chi\left(\chi_i\right) d\chi_i \tag{7.43}$$

where $f_X(\chi_i)$ is the PDF of the random variable χ_i, such that $\chi_i = r_{n_i}^{\varepsilon_i}$ with $r_{n_i} \sim f_R(r_{n_i})$ is obtained as:

$$f_X(\chi_i) = \sum_{l_i}^{L_i} \frac{v_{l_i}}{\varepsilon_i} \chi_i^{\beta_{l_i}-1} \quad 0 < \chi_i \le 1 \tag{7.44}$$

Substituting CDF $F_Y(\cdot)$ from (7.14) and PDF $f_X(\chi_i)$ from (7.44) into (7.43), we obtain

$$F_Z(z_i) = \sum_{j_i=0}^{N_{T_i}} \sum_{l_i}^{L_i} \frac{d_{j_i} v_{l_i}}{\varepsilon_i} \int_0^1 \chi_i^{\beta_{l_i}-1} \frac{\Upsilon(j_i+1, z_i\mu_i\chi_i)}{\Gamma(j_i+1)} d\chi_i \tag{7.45}$$

Now expand the incomplete gamma function in (7.45) using [18, Eq. 8.352.1], we get

$$F_z(z_i) = \sum_{j_i=0}^{N_{T_i}} \sum_{l_i}^{L_i} \frac{d_{j_i} v_{l_i}}{\varepsilon_i} \int_0^1 \chi_i^{\beta_{l_i}-1} d\chi_i - \sum_{j_i=0}^{N_{T_i}} \sum_{l_i}^{L_i} \sum_{k=0}^{j_i} \frac{d_{i_i} v_{l_i}}{\varepsilon_i} \frac{(\mu_i z_i)}{k!} \int_0^1 e^{-z_i\chi_i\mu_i} \chi_i^{k+\beta_{l_i}-1} d\chi_i \tag{7.46}$$

Using area property on PDF function in (7.13) and (7.44), i.e., $\sum_{j_i=0}^{N_{T_i}} d_{j_i} = 1$ and $\sum_{l_i}^{L_i} \frac{v_{l_i}}{\varepsilon_i} \int_0^0 \chi_i^{\beta_{l_i}-1} d\chi_i = 1$ followed by solving integral in (7.46) by [18, Eq. 8.350.1], CDF of Z_i is obtained as

$$F_Z(z_i) = 1 - \sum_{j_i=0}^{N_{T_i}} \sum_{l_i=1}^{L_i} \sum_{k=0}^{j_i} \frac{d_{j_i}}{(k+1)} \frac{v_{l_i}}{\varepsilon_i \mu_i^{\beta_{l_i}}} \frac{\Upsilon(k+\beta_{l_i}, \mu_i z_i)}{z_i^{\beta_{l_i}}} \quad z_i > 0 \tag{7.47}$$

7.4.2 PDF $F_Z(z_i)$ for the random variable Z_i

PDF of Z_i is obtained by differentiating the CDF expression obtained in (7.47) as

$$f_z(z_i) = \frac{d}{dz_i} F_Z(z_i) = \int_0^1 \chi_i f_Y(z_i\chi_i) f_X(\chi_i) d\chi_i \tag{7.48}$$

Now substituting (7.13) and (7.44) into (7.48), we get

$$f_Z(z_i)=\sum_{j_i=0}^{NT_i}\sum_{l_i}^{L_i}\frac{d_{ji}}{\Gamma(j_i+1)}\frac{v_{l_i}}{\varepsilon_i}(\mu_i)^{j_i+1}z_i^{j_i}\int_0^1\chi_i^{j_i+\beta_{l_i}}e^{-z_i\mu_i\chi_i}d\chi_i \tag{7.49}$$

Solving integral in (7.49) with the aid of [18, Eq.8.350.1], corresponding PDF of Z is obtained as:

$$f_z(z_i)=\sum_{j_i=0}^{NT_i}\sum_{l_i=1}^{L_i}\frac{d_{ji}}{\Gamma(j_i+1)}\frac{v_{l_i}}{\varepsilon_i\mu_i^{\beta_{l_i}}}\frac{\Upsilon(j_i+\beta_{l_i}+1,\mu_i z_i)}{z_i^{\beta_{l_i}+1}} \tag{7.50}$$

The given CDF and PDF expressions of the RV are crucial for performance evaluation, particularly in small cell network scenarios, where node mobility leads to substantial variations in received signal power. In this context, the derived statistics effectively capture the impact of small-scale fading. The PDF of Z is analyzed [7] for various mobility topologies of 1D, 2D, and 3D and at path loss of $\varepsilon = 2$. The PDF for the static node scenario depicts the FTR fading channel PDF reported in the literature [16], and the PDF for the mobile scenario under different topologies represents a new derived expression for RWP-enabled FTR fading channel PDF. Here we assumed the unity maximal coverage distance range $R_i = 1m$ and μ_i to be a constant at $1/2$ for both scenarios. The PDF plot [7] for the static case converges early while for the mobile case, the PDF converges slowly depicting the extent of randomness present, caused due to mobility in nodes. The result also depicts that the probability of receiving less power is higher for static nodes compared to mobile nodes and with the increasing topological dimension probability of receiving less power by any node in a specified area reduces.

7.4.3 DF relaying with single antenna receiver

This section derives the SOP expression for DF relay for single antenna receiver. To initially derive the mathematical derivation we normalize $r_{n_i} = r_i / R_i$ and by using (7.18) and (7.19), the received SNR at relay and destination can be expressed as:

$$\gamma_r=\frac{P_{st}|h_1|^2}{r_1^{\varepsilon_1}\sigma_r^2}=\frac{P_{st}|h_1|^2}{R_1^{\varepsilon_1}r_{n1}^{\varepsilon_1}\sigma_r^2}, \tag{7.51}$$

$$\gamma_d(t)=\frac{2\eta\tau P_{st}|h_1|^2|h_2|^2}{(1-\tau)r_1^{\varepsilon_1}r_2^{\varepsilon_2}\sigma_d^2}=\frac{2\eta\tau P_{st}|h_1|^2|h_2|^2}{(1-\tau)R_1^{\varepsilon_1}R_2^{\varepsilon_2}r_{n1}^{\varepsilon_1}r_{n2}^{\varepsilon_2}\sigma_d^2}. \tag{7.52}$$

Now we rewrite (7.2) and by making some rearrangements, Equation (7.2) can be re-formulated. From (7.21), in order to compute $SOP_{DF}^{dyn}(C_{th})$, we need to evaluate the probabilities $P_1(C_{th})$ and $P_2(C_{th})$ which are redefined here for the case of both mobile link scenarios and are derived subsequently. Now by substituting $Z_1 = \frac{|h_1|^2}{r_{n_1}^{\varepsilon_1}}$ into (7.51) and using (7.1), $P_1(C_{th})$ is redefined as:

$$P_1(C_{th}) = \Pr\left(\frac{(1-\tau)}{2}\log_2\left(1+\frac{P_{st}Z_1}{R_1^{\varepsilon_1}\sigma_r^2}\right) \geq C_{th}\right) \tag{7.53}$$

$$= \Pr\left(\frac{P_{st}Z_1}{R_1^{\varepsilon_1}\sigma_r^2} \geq v\right) = \Pr(Z_1 \geq a_1 v) = \bar{F}_{Z_1}(a_1 v)$$

where $v = 2^{\frac{2C_{th}}{1-\tau}} - 1$, $a_1 = \frac{R_1^{\varepsilon_1}\sigma_r^2}{P_{st}}$, and $F_{Z_1}(\cdot)$ is the CDF of random variable Z_1. Now substitute $Z_i = \frac{|h_i|^2}{r_{n_i}^{\varepsilon_i}}$, $i = 1,2$ in (7.52) and using (7.1), $P_2(C_{th})$ is written as:

$$P_2(C_{th}) = \Pr\left(Z_1 \geq a_1 v, \frac{2\eta\tau P_{st}Z_1 Z_2}{R_1^{\varepsilon_1}R_2^{\varepsilon_2}(1-\tau_i)\sigma_d^2} < v\right) \tag{7.54}$$

$$\bar{F}_{Z_1}(a_1 v) - \int_{a_1 v}^{\infty} \bar{F}_{Z_2}\left(\frac{a_2 v}{z}\right) f_{z_1}(z)\,dz$$

From (7.23), (7.53), and (7.54)

$$SOP_{DF}^{dyn}(C_{th}) = 1 - \int_{a_1 v}^{\infty} \bar{F}_{Z_2}\left(\frac{a_2 v}{z}\right) f_{z_1}(z)\,dz \tag{7.55}$$

where $a_2 = \frac{(1-\tau)\sigma_d^2 R_1^{\varepsilon_1} R_2^{\varepsilon_2}}{2\eta\tau P_{st}}$. $SOP_{DF}^{dyn}(C_{th})$ is obtained by substituting PDF, CDF of Z_1 and Z_2 from (7.47), (7.50) into (7.55).

$$SOP_{DF}^{dyn}(C_{th}) = 1 - \sum_{j2=0}^{N_{T2}}\sum_{l_2=1}^{L_2}\sum_{k=0}^{j2}\sum_{j_1=1}^{N_{T1}}\sum_{l_1=1}^{L_1} \frac{d_{j1}d_{j2}}{\Gamma(j_1+1)\Gamma(k+1)} \frac{v_{l_1}v_{l_2}}{\varepsilon_1\varepsilon_2\mu_1^{\beta_{l_1}}\mu_2^{\beta_{l_2}}(a_2 v)^{\beta_{l_2}}}$$

$$\int_{a_1 v}^{\infty} \frac{\Upsilon(j_1+\beta_{l_1}+1, \mu_1 z_1)\Upsilon\left(k+\beta_{l_2}, \mu_2\frac{a_2 v}{z}\right)}{z^{\beta_{l_1}-\beta_{l_2}+1}}dz \tag{7.56}$$

Considering the mobility of the relay and destination node the SOP can be evaluated using the expression obtained in (7.56). The impact of variable parameters on $SOP_{df}^{dyn}(C_{th})$ is analyzed using (7.56) on MATLAB®. However, the variation due to the path loss exponent can be directly deduced from the equation. As ε_i $(i=\{1,2\})$ increases, it increases overall $SOP_{df}^{dyn}(C_{th})$. Note that deriving a closed-form expression for Equation (7.56) is a complex task, as it involves several series approximations that can introduce undesirable inaccuracies. However, the expression consists of readily available functions and can be evaluated using widely accessible mathematical software such as MATLAB [19] or MATHEMATICA [20]. To assist with evaluating (7.56), a MATLAB function has been provided in Appendix D.

7.4.4 AF relaying with single antenna receiver

This section derives the SOP expression for the AF relay-based SWIPT system. The received SNR at the destination node γ_d is re-formulated considering the mobile link behavior of both links as:

$$\gamma_d = \frac{2\eta\tau P_{st}|h_1|^2|h_2|^2}{2\eta\tau r_1^{\varepsilon_1}\sigma_r^2|h_2|^2+(1-\tau)r_1^{\varepsilon_1}r_2^{\varepsilon_2}\sigma_d^2} = \frac{\frac{2\eta\tau P_{st}}{R_1^{\varepsilon_1}R_2^{\varepsilon_2}}Z_1Z_2}{\frac{2\eta\tau\sigma_r^2 Z_2}{R_2^{\varepsilon_2}}+(1+\tau)\sigma_d^2} \tag{7.57}$$

$$= \frac{b_1Z_1Z_2}{b_3Z_2+b_2}$$

where $b_1 = \frac{2\eta\tau P_{st}}{R_1^{\varepsilon_1}R_2^{\varepsilon_2}}$, $(1-\tau)\sigma_d^2$, $b_3 = \frac{2\eta\tau\sigma_r^2}{R_2^{\varepsilon_2}}$, and $Z_i = \frac{|h_i|^2}{r_{n_i}^{\varepsilon_i}}$, $i=1,2$. Now using (7.57), (7.3), and (7.1), SOP for AF relay is given as:

$$SOP_{AF}^{dyn}(C_{th}) = \Pr\left(\frac{b_1Z_1Z_2}{b_3Z_2+b_2}<v\right) = \int_0^\infty F_{Z_1}\left(\frac{vb_3}{b_1}+\frac{b_2v}{b_1z}\right)f_{z2}(z)dz, \tag{7.58}$$

Hence with the aid of (7.47), (7.50) into (7.58) SOP_{AF}^{dyn} is evaluated as:

$$SOP_{AF}^{dyn}(C_{th}) = 1-\sum_{j_1=0}^{N_{T_i}}\sum_{l_1=1}^{L_1}\sum_{j_2=0}^{N_{T_2}}\sum_{l_2=1}^{L_2}\sum_{k=0}^{j_1}\frac{d_{j_1}d_{j_2}}{\Gamma(k+1)\Gamma(j_2+1)}\frac{v_{l_1}v_{l_2}}{\varepsilon_1\varepsilon_2\mu_1^{\beta_{l_1}}\mu_2^{\beta_{l_2}}} \tag{7.59}$$

$$\int_0^\infty \frac{\Upsilon\left(k+\beta_{l_1},\mu_1\left(\frac{vb_3}{b_1}+\frac{b_2v}{b_1z}\right)\right)}{\left(\frac{vb_3}{b_1}+\frac{b_2v}{b_1z}\right)^{\beta_{l_1}}}\frac{\Upsilon(j_2+\beta_{l_2}+1,\mu_2z)}{z^{\beta_{l_2}+1}}dz$$

Thus, the SOP for a mobile relay using the AF protocol with mobile destination nodes can be determined using (7.59). However, deriving a closed-form expression for (7.59) is challenging, as it involves multiple series approximations that may introduce inaccuracies. Nevertheless, the given expression can be efficiently computed using software tools like MATLAB or MATHEMATICA. To facilitate this evaluation, a MATLAB function has been provided in Appendix E.

7.5 RESULTS AND DISCUSSION

The results are presented for Case I and Case II as discussed above. The results are validated with Monte Carlo simulations. The simulation parameters have considered the RWP model as [21] break time of 15 seconds. The selected mobility speeds range from 0 to 10 m/s, with angular motion varying between 0 and 2π being uniformly distributed random variables. The total simulation time is taken to be 200 seconds.

The plots are depicted for different channel scenarios so the results are plotted against different values of k, Δ, and m. The results exhibit the impact of system parameters like the EH time factor τ, source transmits power P_{st}, source-to-relay maximum coverage distance R_1, threshold capacity C_{th}, and diversity branches H. The high correlation between numerical and simulation results depicts the accuracy of the obtained results. Unless otherwise specified we considered the following system parameters similar to the work investigated in [5]. Source transmission power $P_{st} = 1W$, the energy conversion efficiency $\eta = 1$, relay and destination noise power is given by $\sigma_r^2 = \sigma_d^2 = 0.005W$, transmission rate threshold $C_{th} = 0.2$ bps/Hz, source-to-relay link distance is assumed to be $R_1 = r_1 = 6m$.

We assume the instantaneous distance from the relay to the destination as $r_2 = 8m$. Considering the small cell deployment scenario the maximum end-to-end link distance $R_1 + R_2$ is kept at 30 m. EH time factor $\tau = 0.25$, path loss exponent is $\varepsilon_1 = \varepsilon = 2$ and $\varepsilon_2 = 2.5$. In general, we assume the 3D RWP model unless otherwise specified.

7.5.1 Impact of τ and FTR channel parameters on SOP performance for user mobile scenario

The impact of its variation on SOP is summarized in Table 7.3. As normally expected, for high values of k, SOP should be lower with a high availability of LoS power paths. But this is not the case, as Table 7.2 shows that Δ has an important role to play. The reason could be that for high K values the phase alignment of specular components significantly impacts SOP performance. When in phase, the specular components constructively combine, enhancing the overall signal power. Conversely, in the anti-phase condition, the specular waves cancel each other out, leaving only low-powered diffused components, which adversely affect performance. As earlier, the DF relay offers superior outage performance compared to the AF relay with a significant variation seen at higher K and lower Δ value for instance at $\tau = 0.5$ with $k = 10$ and $\Delta = 0.1$, 32.6% improvement is seen. However, for low k and high Δ ($k = 2$ and $\Delta = 0.9$), 12.93% low SOP performance improvement is seen for DF relay compared to its AF counterpart. The reason is that for the favorable channel conditions, DF can perform its decoding operation more efficiently.

Further, it is seen that the selection of the optimal time allocation factor τ also plays an important role in studying the performance variation between AF and DF relaying. As we observe for $\tau = 0.35$ to $\tau = 0.6$ the SOP performance difference between AF and DF relay is more significant as compared to other τ ranges. For low τ, the relay can harvest less energy hence DF relay cannot appropriately perform its decoding operation when inadequate energy is available, hence its performance is comparable to AF. Further for very high τ, ID cannot happen appropriately leading to signal outage for both relay cases. The SOP improves with an increase in H; however, percent improvement in SOP reduces by increasing diversity branches. For instance, the diversity gain observed for the case of without diversity and $H = 2$ is approximately 68% while for $H = 2$ to $H = 3$ it is approx. 43%. Interestingly, it is seen from Table 7.4 that with the increase in H the

Table 7.3 Channel conditions impact on SOP performance

k	Δ	*SOP performance of AF/DF relays*	*SOP variation*
High	High Δ	SOP increases	Significantly high
	Low Δ	SOP decreases	
Low	High Δ	SOP increases	Very insignificant
	Low Δ	SOP increases	

Table 7.4 Impact of channel conditions on AF and DF relay SOP performance

Channel condition			*AF relay*	*DF relay*	*SOP performance variation*
$k = 10$	$\Delta = 0.1$		2.825×10^{-5}	2.25×10^{-5}	**20.3%**
$k = 10$	$\Delta = 0.9$		0.009137	0.008108	11.26%
$k = 2$	$\Delta = 0.1$		0.0577	0.054	6.41%
$k = 2$	$\Delta = 0.9$		0.09758	0.09192	5.8%
$k = 1$	$\Delta = 0.1$	$H = 1$	0.1631	0.1553	**4.78%**
		$H = 2$	0.05977	0.05163	**13.61%**
		$H = 3$	0.03683	0.0289	**21.53%**

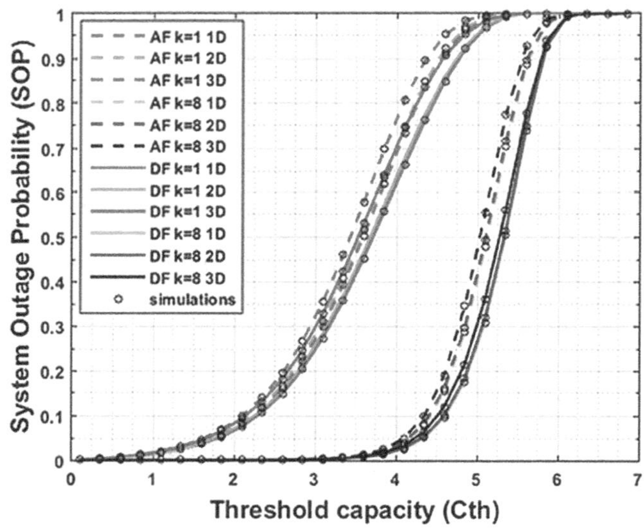

Figure 7.2 SOP as a function of both relay protocol considering 1D, 2D, and 3D topology.

SOP variation in AF and DF relay becomes more prominent. It is observed that with the increase in diversity order, the AF and DF SOP variation is seen for a wide range of τ due to better energy efficiency of the system with diversity order. Further, for favorable channel conditions of high K and low Δ the performance variation in both relays is more significant. The reason can be stated as; the DF relay can perform its decoding process more appropriately when the channel scenario is favorable. Conversely, when the channel is unfavorable, DF processing capability will be inappropriately done. To quantify the above analysis we have presented Table 7.4 calculated for $\tau = 0.3$.

7.5.2 Impact of C_{th} on SOP performance under mobile user scenario

Figure 7.2 depicts the SOP plot as a function of threshold capacity C_{th} for both relaying protocols at $\kappa = \{1,8\}$ assuming 1D, 2D, and 3D mobility topologies. Additionally, this result includes the influence of channel parameter k on the required C_{th} for desired SOP performance.

It is seen from the results that for higher k better SOP performance can be obtained even for high C_{th} value. The explanation for the same can be illustrated as in high k (and low Δ) average power associated with specular component increases, which enhances the channel conditions and hence signal reception improves. Further, for extremely high or extremely low C_{th}

values the variation is negligible because for low C_{th} the required data rate is very low which is met by all the considered cases even in the worst channel condition hence outage is very low.

7.5.3 SOP as a function of τ for static and mobile cases

To demonstrate the comparison of both static and mobile cases, we have presented a comprehensive plot of SOP in terms of the EH factor τ. It is mentioned here that the plot corresponding to the static node scenario is presented in [14] that considers a cooperative SWIPT system with AF and DF relay over the FTR fading channel.

In this chapter, we extend the analysis presented in [14] by exploring node mobility and its impact due to (i) user mobile terminal and (ii) both relay and user mobile terminal, which is inherent in wireless communication systems but has been ignored in most of the studies. To make the result tractable under this subsection, the results are simulated with the parameters as in [14] with $P_{st} = 20W$, $r_1 = r_2 = 3m$, $\sigma^2 = 1$, $C_{th} = 0.1$ bit/s/Hz and $H = 1$ to aid in evaluation.

The result in Figure 7.3 provides the plot for SOP versus τ under static and mobile scenarios for both relaying protocols. It is observed here that the SOP improves under the mobile node scenario compared to the static counterpart. It is due to the fact that for the case of random link distance,

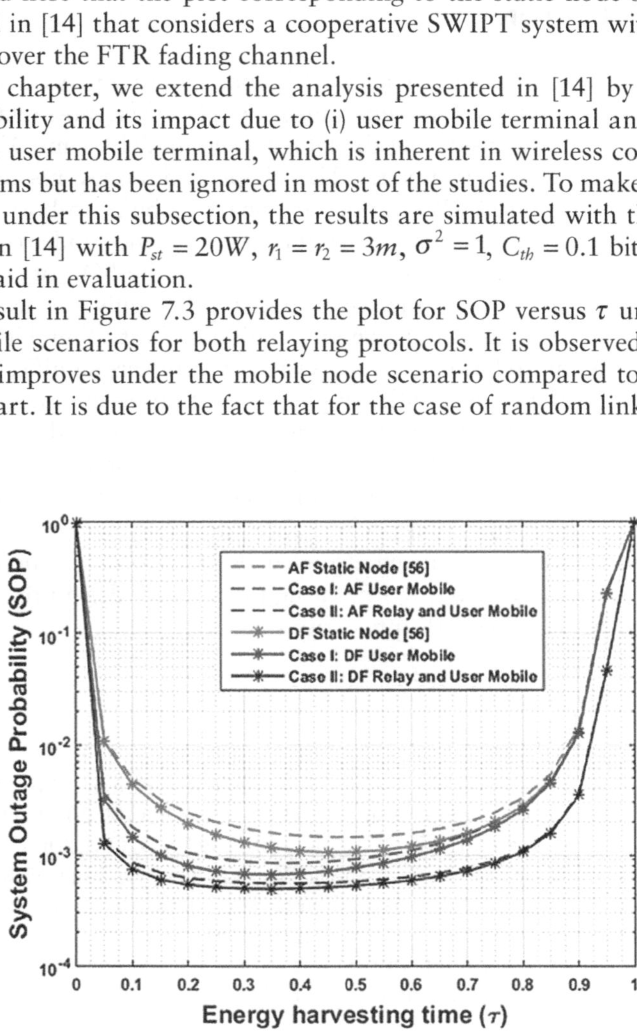

Figure 7.3 SOP vs plot for AF and DF relay.

outage performance is evaluated by averaging over the entire radial transmission range; hence the probability of outage at a particular instant is improved. Further, the performance gap between AF and DF relay reduces under mobile scenario compared to static counterpart. Also, it is seen from the result that optimal τ^* range becomes more skewed for the static case; however, with consideration of mobility, superior outage performance can be achieved for a wide range of τ^* demonstrating the flexibility in design with the mobile nodes.

7.6 CONCLUSION

In the realm of UAV-based IoT networks, node mobility plays a critical role in ensuring seamless wireless communication. However, most relay-enabled IoT SWIPT studies considering the FTR fading model for mmWave bands have largely overlooked its impact. This chapter addresses this gap by integrating the widely used RWP model to impart mobility to both relay and IoT sensor nodes, thereby analyzing outage performance under two distinct scenarios. We derive closed-form expressions for the SOP of AF and DF relaying protocols, considering cases where either only the user node is mobile or both the relay and user nodes exhibit mobility. Simulation results reveal that the interrelation between system parameters significantly influences outage performance, with a higher impact observed under favorable channel conditions. Furthermore, we establish that the trade-off between the EH time factor and diversity, previously studied for static nodes, remains valid in mobile scenarios but exhibits varying degrees of SOP improvement. The study also highlights that performance variations between AF and DF relays become more pronounced at higher diversity orders. These findings underscore the necessity of incorporating mobility models in relay-assisted SWIPT systems for UAV-IoT applications.

REFERENCES

1. Lin, Guolong, Guevara Noubir, and Rajmohan Rajaraman. "Mobility models for ad hoc network simulation." In *IEEE INFOCOM 2004*, Hong Kong, China, vol. 1, p. 463. IEEE, 2004. doi: 10.1109/INFCOM.2004.1354517.
2. Zhao, Nan, Weidang Lu, Min Sheng, Yunfei Chen, Jie Tang, F. Richard Yu, and Kai-Kit Wong. "UAV-assisted emergency networks in disasters." *IEEE Wireless Communications* 26, no. 1 (2019): 45–51.
3. Li, Bin, Zesong Fei, and Yan Zhang. "UAV communications for 5G and beyond: Recent advances and future trends." *IEEE Internet of Things Journal* 6, no. 2 (2018): 2241–2263.
4. Zeng, Yong, and Rui Zhang. "Energy-efficient UAV communication with trajectory optimization." *IEEE Transactions on Wireless Communications* 16, no. 6 (2017): 3747–3760.

5. Rabie, Khaled Maaiuf, Bamidele Adebisi, and Mohamed-Slim Alouini. "Half-duplex and full-duplex AF and DF relaying with energy-harvesting in log-normal fading." *IEEE Transactions on Green Communications and Networking* 1, no. 4 (2017): 468–480.
6. Singh, Shweta, Debjani Mitra, and Rajendra Kumar Baghel. Impact of user mobility on wireless powered network for AF and DF relaying over FTR fading channel. *Wireless Networks* 27 (2021 Nov): 5057–5072.
7. Singh, Shweta, Debjani Mitra, and Rajendra Kumar Baghel. Co-operative SWIPT system over FTR fading channel in mobile network scenario. *Journal of Ambient Intelligence and Humanized Computing* 14, no. 8 (2023 Aug): 11139–11152.
8. Hyytiä, Esa, and Jorma Virtamo. "Random waypoint mobility model in cellular networks." *Wireless Networks* 13, no. 2 (2007): 177–188.
9. Bettstetter, Christian, Giovanni Resta, and Paolo Santi. "The node distribution of the random waypoint mobility model for wireless ad hoc networks." *IEEE Transactions on Mobile Computing* 2, no. 3 (2003): 257–269.
10. Hyytia, Esa, Pasi Lassila, and Jorma Virtamo. "Spatial node distribution of the random waypoint mobility model with applications." *IEEE Transactions on Mobile Computing* 5, no. 6 (2006): 680–694.
11. Li, Chunxiao, Jie Yao, Haitao Wang, Usman Ahmed, and Shixiao Du. "Effect of mobile wireless on outage and BER performances over Rician fading channel." *IEEE Access* 8 (2020): 91799–91806.
12. Hyytiä, Esa, and Jorma Virtamo. "Random waypoint model in *n*-dimensional space." *Operations Research Letters* 33, no. 6 (2005): 567–571.
13. Badarneh, Osamah S., Daniel Benevides Da Costa, and Pedro Henrique Juliano Nardelli. "Wireless-powered communication networks with random mobility." *IEEE Access* 7 (2019): 166476–166492.
14. Zheng, Jiakang, Jiayi Zhang, Shuaifei Chen, Hui Zhao, and Bo Ai. "Wireless powered UAV relay communications over fluctuating two-ray fading channels." *Physical Communication* 35 (2019): 100724.
15. Govindan, Kannan, Kai Zeng, and Prasant Mohapatra. "Probability density of the received power in mobile networks." *IEEE Transactions on Wireless Communications* 10, no. 11 (2011): 3613–3619.
16. Zhang, Jiayi, Wen Zeng, Xu Li, Qiang Sun, and Kostas P. Peppas. "New results on the fluctuating two-ray model with arbitrary fading parameters and its applications." *IEEE Transactions on Vehicular Technology* 67, no. 3 (2017): 2766–2770.
17. Shah, Manilal. "On generalizations of some results and their applications." *Collectanea Mathematica* 24 (1973): 249–266.
18. Gradshteyn, Iosif Moiseevich, and Iosif Matveevich Ryzhik. *Table of Integrals, Series, and Products*, edited by a. New York: Jeffrey Academic; 2000. p. 658.
19. MATLAB. 9.7.0.1190202 *(*R2015a*)*. Natick, MA: The MathWorks Inc.; 2015.
20. Wolfram Research, Inc. *Mathematica, Version 12.3*. Champaign, IL; 2021.
21. Appiah, Martin. "The impact of mobility models on the performance of mobile ad hoc network (MANET)." In *2016 International Conference on Advances in Computing and Communication Engineering (ICACCE)*, Durban, South Africa, pp. 208–213. IEEE, 2016. doi: 10.1109/ICACCE.2016.8073749.

APPENDIX

Appendix A Closed form for integral I_1

A generalized form for integral I_1 in (7.27) is given by (7.60)

$$I_1 = \int_{\chi}^{\infty} z^{j_1} e^{-\alpha z} \Upsilon\left(j_2 + 1, \frac{\beta}{z} \right) dz \tag{7.60}$$

where we assume $\chi = a_1 v$, $\alpha = 1/2\sigma_1^2$, $\beta = \frac{a_2 v}{2\sigma_2^2}$. Expand incomplete gamma function in (3.36) and using [18, 8.352.6]

$$I_1 = \Gamma(j_2 + 1)\left[\int_{\chi}^{\infty} z^{j_1} e^{-\alpha z} dz - \sum_{n=0}^{[j_2]} \frac{\beta^n}{n!} \int_{\chi}^{\infty} z^{j_1 - n} e^{-\alpha z - \frac{\beta}{z}} dz \right] \tag{7.61}$$

With the aid of [18, 3.351.2] and [20] a closed-form integral is obtained in (3.38)

$$I_1 = \frac{\Gamma(j_2 + 1)\Gamma(j_1 + 1, \chi)}{\alpha^{j_1 + 1}} - \Gamma(j_2 + 1)\chi^{j_1 + 1} \sum_{n=0}^{j_2} \frac{1}{n!}\left(\frac{\beta}{\chi} \right)^n \tag{7.62}$$

$$G_{1,1;1,0;0,1}^{1,0;0,1;1,0}\left(\begin{matrix} n - j_1 \\ n - j_1 + 1 \end{matrix} \middle| \begin{matrix} 1 \\ - \end{matrix} \middle| \begin{matrix} - \\ 0 \end{matrix} \middle\| \frac{1}{\chi\alpha}, \frac{\beta}{\chi} \right)$$

where $G_{\cdot,\cdot;\cdot,\cdot;\cdot,\cdot}^{\cdot,\cdot;\cdot,\cdot;\cdot,\cdot}(\cdot|\cdot|\cdot|\cdot,\cdot)$ represents EGBMGF [20].

Appendix B Closed form for integral I_7

Generalized form for integral I_7 in (7.30) can be represented as

$$I_7 = \int_0^{R_2} r_2^{\rho} \int_{\chi}^{\infty} z^{j_1} e^{-\alpha z} \Upsilon\left(j_2 + 1, \beta \frac{r_2^{\varepsilon}}{z} \right) dz dr_2 \tag{7.63}$$

Changing order of integral and using transformation $r_2^{\varepsilon} = R_2^{\varepsilon} x$, we get

$$I_7 = \frac{R_2^{\rho+1}}{\varepsilon} \int_{\chi}^{\infty} z^{j_1} e^{-\alpha z} \int_0^1 x^{w-1} \Upsilon\left(j_2 + 1, \frac{\beta R_2^{\varepsilon}}{z} x \right) dx dz \tag{7.64}$$

where $w = (\rho + 1)/\varepsilon$

Using integration by parts it can be shown that

$$\int_0^1 z^{w-1}\Upsilon(m,ax)dx = w^{-1}\left[\Upsilon(m,a)-a^{-w}\Upsilon(w+m,a)\right] \tag{7.65}$$

Hence from (4.64) and (4.65),

$$I_7 = \frac{R_2^{\rho+1}}{\varepsilon w}\underbrace{\int_0^\infty z^{j_1}e^{-\alpha z}\Upsilon\left(j_2+1,\frac{\beta R_2^\varepsilon}{z}\right)dz}_{I_9}-\left(\beta R_2^\varepsilon\right)^{-w}\frac{R_2^{\rho+1}}{\varepsilon w}\underbrace{\int_0^\infty z^{j_1+w}e^{-\alpha z}\Upsilon\left(w+j_2+1,\frac{\beta R_2^\varepsilon}{z}\right)dz}_{I_{10}}. \tag{7.66}$$

A closed-form expression for integral I_7 can be found by solving I_9 and I_{10}. Let's define integral I_9 as function $f\left(j_1,\lceil j_2+1\rceil\right)$ defined as follows:

$$f\left(j_1,\lceil j_2+1\rceil\right)=\int_\chi^\infty z^{j_1}e^{-\alpha z}\Upsilon\left(j_2+1,\frac{\beta {R_2}^\varepsilon}{z}\right)dz \tag{7.67}$$

Expand incomplete gamma function as in [18, 8.352.6]

$$f\left(j_1,\lceil j_2+1\rceil\right)=\Gamma\left(j_2+1\right)\left[\int_\chi^\infty z^{j_1}e^{-\alpha z}dz-\sum_{n=0}^{\lceil j_2\rceil}\frac{\left(\beta R_2^\varepsilon\right)^n}{n!}\int_\chi^\infty z^{j_1-n}e^{-\alpha z-\frac{\beta R_2^\varepsilon}{z}}dz\right] \tag{7.68}$$

Using [18, 3.351.2] and [20] a closed form for integrals is given as

$$f\left(j_1,\lceil j_2+1\rceil\right)=\frac{\Gamma\left(j_2+1\right)\Gamma\left(j_1+1,\chi\right)}{\alpha^{j_1+1}}-\Gamma\left(j_2+1\right)\chi^{j_1+1}\sum_{n=0}^{\lceil j_2\rceil}\frac{1}{n!}\left(\frac{\beta R_2^\varepsilon}{\chi}\right)^n \tag{7.69}$$

$$G_{1,1;1,0;0,1}^{1,0;0,1;1,0}\left(\begin{array}{c|c|c|c} n-j_1 & 1 & - & \\ n-j_1+1 & - & 0 & \end{array}\frac{1}{\chi\alpha},\frac{\beta R_2^\varepsilon}{\chi}\right)$$

where $G_{\cdot,\cdot;\cdot,\cdot;\cdot,\cdot}^{\cdot,\cdot;\cdot,\cdot;\cdot,\cdot}\left(\cdot|\cdot|\cdot|\cdot,\cdot\right)$ represents EGBMGF [20]. Now using (4.72) and (4.75),

$$I_7=\frac{R_2^{\rho+1}}{\varepsilon w}\left(f\left(j_1,\lceil j_2+1\rceil\right)-\left(\beta R_2^\varepsilon\right)^{-w}f\left(j_1+w,\lceil j_2+w+1\rceil\right)\right) \tag{4.76}$$

Appendix C Closed form for integral I_8

A generalized form for integral I_8 in (7.35) is given as

$$I_8=\int_0^{R_2} r_2^\rho\int_0^\infty z^{j_2}e^{-\alpha z}\Upsilon\left(j_1+1,\left(\beta+\chi\frac{r_2^\varepsilon}{z}\right)\right)dzdr_2. \tag{7.71}$$

Use [18, 8.352.6] to expand incomplete gamma function

$$\Upsilon\left(j_1+1,\left(\beta+\chi\frac{r_2^{\varepsilon}}{z}\right)\right)=\Gamma(j_1+1)\left(1-e^{-\left(\beta+\chi\frac{r_2^{\varepsilon}}{z}\right)}\sum_{n=0}^{j_1}\frac{1}{n!}\left(\beta+\chi\frac{r_2^{\varepsilon}}{z}\right)^{n}\right) \quad (7.72)$$

Using binomial expansion and re-arranging (4.78), we get

$$\Upsilon\left(j_1+1,\left(\beta+\chi\frac{r_2^{\varepsilon}}{z}\right)\right)=\Gamma(j_1+1)$$

$$\left(1-e^{-\left(\beta+\chi\frac{r_2^{\varepsilon}}{z}\right)}\sum_{n=0}^{j_1}\sum_{m=0}^{n}\binom{n}{m}\frac{\beta^{n-m}\chi^{n}}{n!}\left(\frac{r_2^{\varepsilon}}{z}\right)^{n}\right). \quad (7.73)$$

From (4.77) and (4.79),

$$I_8=\Gamma(j_1+1)\underbrace{\int_0^{R_2}r_2^{\rho}\int_0^{\infty}z^{j_2}e^{-\alpha z}dz dr_2}_{I_{11}}-e^{-\beta}\Gamma(j_1+1) \quad (7.74)$$

$$\sum_{n=0}^{j_1}\sum_{m=0}^{n}\binom{n}{m}\frac{\beta^{n-m}\chi^{n}}{n!}\underbrace{\int_0^{R_2}r^{\rho+\varepsilon n}\int_0^{\infty}z^{j_2-n}e^{-\left(\alpha z+\frac{X^{r^{\varepsilon}}}{z}\right)}dz\, dr_2}_{I_{12}}.$$

Integral I_{11} in (4.80) is solved using [18, 3.351.3], where I_{12} can be given as a modified Bessel function of second kind using [18, 3.471.9]. In terms of Meijer-G function using [18, 9.34.3], we get

$$I_8=\Gamma(j_1+1)\frac{R_2^{\rho+1}}{\rho+1}\frac{\Gamma(j_2+1)}{\alpha^{j_2+1}}-e^{-\beta}\Gamma(j_1+1)\sum_{n=0}^{j_1}\sum_{m=0}^{n}\binom{n}{m}\frac{\beta^{n-m}\chi^{n}}{n!\alpha^{\omega}} \quad (7.75)$$

$$\int_0^{R_2}r_2^{\rho+\varepsilon n}G_{0,2}^{2,0}\left(\chi r_2^{\varepsilon}\alpha\left|\begin{matrix}-,-\\0,\omega\end{matrix}\right.\right)dr_2$$

where ω is the nearest integer to fraction (j_2-n+1). Using [20], a closed-form expression for integral I_8 is given as

$$I_8 = \Gamma(j_1+1)\frac{R_2^{\rho+1}}{\rho+1}\frac{\Gamma(j_2+1)}{\alpha^{j_2+1}} - e^{-\beta}\Gamma(j_1+1)$$

$$\sum_{n=0}^{j_1}\sum_{m=0}^{n}\binom{n}{m}\frac{\beta^{n-m}\chi^{n-c_l}}{n!\varepsilon\alpha^{\omega+c_l}}G_{1,3}^{2,1}\left(\alpha\chi R_2^{\varepsilon}\middle| \begin{matrix} 1 \\ c_l, c_l+\omega, 0 \end{matrix}\right). \tag{7.76}$$

where $c_l = (\rho_l + m\varepsilon + 1)/\varepsilon$

Appendix D for DF relay

$$I_{DF} = \int_{lo\_lim}^{\infty} \frac{\Upsilon(a_1, b_1 z)}{z^{c_1}} \frac{\Upsilon(a_2, b_2/z)}{(c_2/z)^{c_3}} dz$$

```
function [q] = int_df(a1,a2,b1,b2,c1,c2,c3,lo_lim)
fun = @(z) (gammainc(b1.*z,a1).*gammainc((b2./z),a2))./
((z.^c1).*(c2./z).^c3);
q = gamma(a1)*gamma(a2)*integral(fun,lo_lim,inf);
end
```

Appendix E for AF relay

$$I_{AF} = \int_{0}^{\infty} \frac{\Upsilon(a_1, b_1 z)\Upsilon(a_2, b_2 + b_3/z)}{z^{c_1}(c_2 + c_3/z)^{c_4}} dz$$

```
function [q] = gma_int(a1,a2,b1,b2,b3,c1,c2,c3,c4)
fun = @(z) (gammainc(b1.*z,a1).*gammainc((b2+b3./z),a2))./
((z.^c1).*(c2+c3./z).^c4);
q = gamma(a1)*gamma(a2)*integral(fun,0,inf);
end
```

Chapter 8

Autonomous agricultural robotics and IoT in precision farming

Enhancing productivity through smart automation

Venkata Naga Rani Bandaru, Idamakanti Kasireddy, G. Ratnakanth, Sanapathi Venkata Ashok Durga Prasad, and M. Srikanth

8.1 INTRODUCTION

The foundation of human civilization, agriculture offers millions of people worldwide economic stability, raw materials, and food security. By 2050, there will likely be 9.7 billion people on the planet, and in order to meet this demand, food production would need to increase by almost 70%. However, there are several obstacles facing modern agriculture, such as decreased soil fertility, ineffective resource management, workforce shortages, and climate change. Technological advancements like the Internet of Things (IoT) and autonomous agricultural robotics (AAR) are showing promise as game-changing remedies to these issues. By increasing productivity, precision, and sustainability while reducing the need for human intervention, these technologies have the potential to completely transform conventional farming. Critical farming tasks including seeding, weeding, irrigation, fertilizing, and harvesting can be automated by autonomous agricultural robots that are outfitted with IoT sensors, artificial intelligence (AI), and machine vision. Robotic farming solutions provide real-time decision-making, remote monitoring, and precision-based treatments, in contrast to traditional techniques that depend on physical labor. By linking agricultural machinery, drones, sensors, and databases, the IoT plays a crucial part in giving farmers access to real-time information on crop growth, soil health, weather, and pest infestations. The creation of smart farms that maximize output and resource use is made possible by the smooth integration of IoT with agricultural robotics.

8.1.1 The need for smart automation in agriculture

Conventional farming is time-consuming, labor-intensive, and vulnerable to environmental fluctuations. Farmers have had to contend with erratic weather patterns, degraded soil, ineffective irrigation, and excessive use of

 DOI: 10.1201/9781003589730-8

chemical pesticides and fertilizers throughout the years. Low crop yields, monetary losses, and unsustainable farming methods are all caused by these problems. Automated solutions that can increase productivity while lowering agriculture's ecological footprint are desperately needed, as worries about food security and environmental sustainability grow.

Obstacles in Conventional Agriculture: Labor Shortages: Due to rural-to-urban migration, aging farmer populations, and waning youth interest in agriculture, many agricultural economies struggle with a shortage of competent farm labor.

Climate Change: Crop yields and profitability are impacted by the disruption of farming cycles caused by rising temperatures, unpredictable rainfall, and harsh weather.

Resource Inefficiencies: Soil deterioration, water scarcity, and environmental contamination are caused by excessive use of water, fertilizers, and pesticides.

Management of Pests and Diseases: Manually identifying pests and plant diseases takes a lot of effort and frequently results in postponed interventions, which can result in large crop losses.

Operational expenditures: Wages, machinery fuel, and maintenance expenditures are only a few of the high operational costs associated with traditional farming practices. IoT-enabled AAR provides a sustainable solution to these problems by using automation, real-time data, and AI-driven analytics in precision farming methods.

8.1.2 The role of IoT in AAR

The automation of contemporary agriculture is based on the IoT. Farmers no longer need to be physically present in the fields to remotely monitor and operate a variety of agricultural operations thanks to the integration of IoT and robotics. Drones, cloud computing platforms, Global Positioning System (GPS) modules, and IoT sensors collaborate to gather, process, and evaluate data in real time.

Important Elements of Agricultural Robotics Driven by IoT: Intelligent Sensors: By measuring the amount of water in the soil, soil moisture sensors help automated irrigation systems deliver water effectively.

Nutrient Sensors: Examine the soil's nutrient composition and recommend the best way to apply fertilizer.

Weather Sensors: To support climate adaption plans, track temperature, humidity, wind speed, and atmospheric pressure.

Unmanned Aerial Vehicles (UAVs) and Drones: Farmers can use UAVs with high-resolution cameras, Light Detection and Ranging (LiDAR), and multispectral photography to identify pest infestations, plant health problems, and agricultural diseases from the air. Precision fertilizer and pesticide spraying with drones can reduce chemical waste and its negative effects on the environment.

Robots Driven by AI for Accurate Farming: AI-based weeding equipment, robotic harvesters, and autonomous tractors can work around the clock, lowering the need for human labor. Robots with machine vision capabilities can distinguish between crops and weeds, allowing for focused herbicide delivery rather than careless spraying.

Big Data Analytics with Cloud Computing: IoT devices constantly send data to cloud platforms, where AI algorithms analyze massive datasets to produce insights that can be put to use. Farmers may make proactive decisions by using predictive analytics to forecast production, crop diseases, and weather conditions.

Real-Time Processing and Edge Computing: By ensuring that sensor data is handled locally on the device, edge computing lowers latency and permits real-time reactions. For autonomous robotic systems that need to make decisions instantly, such as self-navigating tractors and real-time pest control systems, this is essential.

8.1.3 Advantages of AAR and IoT

8.1.3.1 Using precision farming to increase output

Highly accurate farming tasks are carried out by automated robots, which save resources and boost crop output per hectare. The proper amount of water, fertilizer, and pesticide is sprayed to the crops thanks to AI-driven decision-making.

8.1.3.2 Efficiency in operations and cost reduction

By automating labor-intensive and repetitive processes like sowing, spraying, and harvesting, robotics lowers labor costs. Operational costs are reduced by fuel-efficient autonomous tractors and drones.

8.1.3.3 Environmental benefits and sustainability

By minimizing chemical runoff into water bodies, targeted pesticide spraying lowers contamination of the soil and water. Water utilization is optimized via automated irrigation systems, which also minimize waste and avoid over-irrigation.

8.1.3.4 Remote control and monitoring

IoT-enabled smartphone applications allow farmers to remotely monitor their whole farm, eliminating the need for in-person field trips. Farmers are informed about disease outbreaks, weather patterns, and irrigation requirements by AI-powered alerts and notifications.

8.1.3.5 Customization and scalability

From small-scale organic farms to massive commercial agribusinesses, IoT-based robotic solutions may be tailored for farms of all sizes. Agricultural robots can be tailored by farmers according to crop varieties, soil types, and geographic location.

8.1.3.6 Challenges and future directions

IoT-integrated AAR has many benefits, but it also has drawbacks. Hardware, software, and infrastructure must be purchased in advance for advanced robotic farming solutions. As the IoT becomes more prevalent, agricultural data security and defense against online attacks become crucial issues (Bandaru et al., 2024a). Farmers and other agricultural professionals must receive technical training in order to adopt smart farming technologies. Real-time IoT data transmission is hampered by rural areas' frequently poor internet connectivity.

By making farming more productive, sustainable, and efficient, the combination of IoT and AAR has the potential to completely transform the agricultural landscape. Farmers can now maximize resources, boost crop yields, and lessen their impact on the environment thanks to these advancements. Cost, cybersecurity, and connectivity issues still exist, but ongoing developments in AI, machine learning (ML), and IoT connectivity should hasten the implementation of smart farming. Intelligent, data-driven, and automated farming systems that guarantee food security and sustainability for future generations are the way of the future.

8.2 THE ROLE OF AAR AND IoT IN TRANSFORMING PRECISION FARMING

Over the past 10 years, a lot of study has been done on the potential benefits of integrating AAR with the IoT, with an emphasis on how it might improve sustainability, resource efficiency, and precision farming. Numerous researches have looked into how to optimize agricultural processes using robotics, AI-driven automation, IoT-enabled sensors, and cloud computing. The previous research contributions in autonomous robotic farming, IoT-based smart agriculture, AI-driven crop monitoring, precision irrigation, automated pest control, and predictive analytics for sustainable agriculture are all thoroughly examined in this literature review.

1. Self-Governing Agricultural Robots for Intelligent Farming: To help with crop cultivation, weeding, harvesting, and irrigation, autonomous robots have been designed. In order to detect plant growth stages and automate harvesting, Bac et al. (2014) conducted early

research on creating self-navigating robots for greenhouse conditions using LiDAR sensors, GPS tracking, and machine vision. Similarly, Duckett et al. (2018) reduced pesticide use and increased yield efficiency by introducing field robots with deep learning models for weed detection. These models have been extended in recent research, which build on foundational work in agricultural field robotics, which set out the concepts and components of robotic systems for tasks such as spraying and crop handling (Bechar & Vigneault, 2016). The development of autonomous farming has been further expedited by the introduction of robotic harvesting. Arad et al. (2021) presented a deep learning-based robotic harvester that uses AI algorithms and hyperspectral imagery to determine the ripeness of fruit. When compared to conventional methods, their study showed a 20% increase in harvesting efficiency. Ju et al. (2022) reviewed multi-robot systems that cooperate on large-scale tasks such as fertilizing, sowing, and plowing. These studies demonstrate how autonomous robots are revolutionizing conventional farming by enhancing precision-based decision-making and lowering reliance on labor.

2. Real-Time Monitoring and Smart Agriculture Powered by IoT: Through the integration of cloud computing, AI-driven analytics, and wireless sensor networks (WSNs), IoT has greatly improved real-time monitoring and decision-making in agriculture. In order to dynamically control the water supply, Patil et al. (2016) presented an IoT-based smart irrigation system that included temperature sensors, soil moisture sensors, and weather forecasting algorithms. Their findings showed a 12% improvement in crop productivity and a 30% decrease in water waste. Other researchers have studied IoT's usefulness in illness prediction and pest control. IoT-based sensing of leaf, temperature, and humidity data supports early detection of crop infections (Ayaz et al., 2019). Furthermore, IoT's effectiveness in smart agriculture has increased because of 5G and edge computing. Low-latency 5G networks' application in remote-controlled tractors and UAV-based agricultural health monitoring was investigated by Jain et al. (2021). Their research found that edge computing-enabled real-time data processing accelerated decision-making by 65%, enabling farmers to respond quickly to pest infestations or changes in the climate. These results show that real-time, precision-based agricultural techniques depend on IoT-driven smart farming.
3. AI-Powered Predictive Analytics and Crop Monitoring: Crop health evaluation, yield prediction, and resource management have all benefited greatly from AI. Deep learning has been applied successfully to image-based crop and disease analysis (Mohanty et al., 2016). More precise crop recognition and yield estimation are now possible

because of developments in convolutional neural networks (CNNs) and ML. CNNs have achieved high accuracy in classifying crop species, growth stages, and pest damage (Kamilaris & Prenafeta-Boldú, 2018). With this degree of accuracy, farmers can reduce crop loss, enhance disease resistance tactics, and use resources as efficiently as possible. To improve farming results, recent research has combined IoT with AI-powered predictive analytics. Machine learning has been integrated across smart-farming tasks including soil analysis and yield estimation (Liakos et al., 2018). In contemporary agriculture, the incorporation of AI and big data analytics into autonomous agricultural systems is revolutionizing the field.

4. Automatic Water Management and Accurate Irrigation: One of the most urgent issues facing agriculture is water scarcity, and precision irrigation methods based on the IoT have shown to be successful remedies. Mandal et al. (2018) created an AI-powered irrigation control system that controlled water distribution by using weather forecasts, evapotranspiration rates, and soil moisture data. According to their findings, they were able to reduce water usage by 45% without sacrificing crop yield. The function of cloud computing in automated irrigation has been examined in several publications. In 2020, Kumar et al. presented a cloud-based irrigation management system that integrated AI-powered water distribution algorithms with data from IoT sensors. Real-time water allocation optimization using the system improved crop hydration levels and cut irrigation expenditures by 28%. These results demonstrate how intelligent irrigation systems can help alleviate water scarcity and promote sustainable agriculture.
5. AAR: Obstacles and Prospects: Even though IoT-enabled agricultural robots have advanced significantly, experts recognize a number of obstacles that prevent widespread adoption. High upfront investment prices, interoperability problems, cybersecurity threats, and a lack of technical know-how were noted by Farooq et al. (2020) as key barriers to adoption. There is a digital divide in agriculture since many small-scale farmers in poor nations do not have access to high-speed internet or robotic systems driven by AI. Emerging research is concentrating on decentralized AI models and affordable robots to address these issues. Swarm robotics, in which several robots cooperate on field activities, has been reviewed as a promising low-cost approach (Albiero et al., 2022). In order to protect agricultural data, stop cyberattacks, and guarantee transparency in AI-driven farming decisions, academics are also investigating blockchain technology.

The literature review highlights how AAR and the IoT are revolutionizing modern farming. Crop productivity, resource efficiency, and sustainability have all been greatly enhanced by smart agricultural technology, such as AI-driven robotic systems, IoT-enabled real-time monitoring, predictive analytics, and precision irrigation. To guarantee the broad adoption of autonomous farming solutions, however, issues like exorbitant prices, cybersecurity threats, and restricted connectivity in rural areas must be resolved. For smart agriculture to be implemented globally, future research should concentrate on robotic frameworks that are AI-driven, scalable, and reasonably priced.

8.3 METHODOLOGY

This research paper's technique, AAR and IoT in Precision Farming, aims to methodically investigate, deploy, and assess smart agricultural systems that combine robotics, IoT, AI, machine vision, and real-time analytics. System design, data collecting, robotic automation, IoT integration, ML algorithms, and evaluation metrics are all part of the methodology's structured approach. The research framework, implementation strategies, and evaluation procedures for determining the effectiveness and impact of autonomous robotic farming solutions are broken down in detail in this part.

8.3.1 Framework for research

The study employs a multi-tiered system that combines cloud computing, robotics, IoT, and AI-driven analytics. The main objective is to build and create an autonomous agricultural system that can optimize resource use and yield efficiency while carrying out tasks including automated harvesting, precision irrigation, real-time crop monitoring, and pest control. There are four main stages to the research framework: System architecture development is the process of creating the software and hardware elements of self-sufficient robotic agricultural systems.

IoT and Sensor Integration: Use smart sensors and IoT devices to implement real-time environmental monitoring. Creating deep learning models for predictive analytics, weed identification, and agricultural health analysis is an example of integrating AI with ML.

Evaluation and Performance Analysis: Using a variety of experimental trials and performance measures, this study evaluates the efficacy of robotic automation and IoT-driven precision farming.

8.3.2 AAR system architecture

The following essential elements make up the system architecture:

8.3.2.1 Self-governing actuators and robots

Drones, automated actuators for carrying out farm tasks, and robotic farming equipment make up the heart of AAR. The purpose of these robots is: GPS and LiDAR are used for autonomous navigation; automated spraying of pesticides, fertilizer, and seeds; robotic arms for plant maintenance and harvesting. To identify crops, detect illnesses, and oversee agricultural operations, robots are outfitted with machine vision cameras, multispectral sensors, and AI-based decision-making algorithms.

8.3.2.2 Data collection and IoT sensor networks

To track current environmental conditions, an IoT sensor network is set up throughout the agricultural area. These consist of: sensors of soil moisture for targeted watering; sensors for nutrients to assess the health of the soil; sensors for tracking the weather to forecast changes in the climate; drones equipped with multispectral imaging cameras to evaluate the health of plants. A cloud-based system receives the gathered sensor data, which is then processed by AI models to produce useful insights.

8.3.2.3 ML models and AI-based data processing

AI is essential for processing and evaluating vast amounts of agricultural data. This study makes use of: using deep learning (CNNs, Recurrent Neural Network (RNNs)) to predict yield, detect diseases, and classify crops; predictive analytics to forecast pest outbreaks and the weather; optimizing robotic decision-making in autonomous farming through reinforcement learning; edge computing for low-latency IoT devices to process data in real time. The AI models are trained using datasets from satellite photos, agricultural research stations, and actual farms to improve accuracy and precision.

8.3.2.4 IoT-based decision support systems and cloud computing

IoT-driven cloud computing systems are integrated in the study for: processing and storing data related to agriculture; provide farmers with real-time dashboards to track crop conditions; allowing farm operations to be automated and controlled remotely. Scalable and real-time data management is accomplished using cloud platforms like Microsoft Azure FarmBeats, Google Cloud Agriculture, and AWS IoT.

Input: Pseudocode for System Architecture

1. Initialize the autonomous agricultural robotic system

   Initialize_Robot()

   Initialize_IoT_Sensors()

   Initialize_Cloud_Server()

   Initialize_AI_Model()

2. Start Data Collection from IoT Sensors

   while True:

   soil_moisture = Read_Sensor("Soil Moisture")

   temperature = Read_Sensor("Temperature")

   humidity = Read_Sensor("Humidity")

   plant_health = Read_Sensor("Multispectral Camera")

3. Send Data to Cloud Server for Processing

   Send_To_Cloud(soil_moisture, temperature, humidity, plant_health)

4. AI Model Predicts Crop Health

   crop_health_status = AI_Model_Predict(plant_health)

5. Decision-Making for Farm Automation

   if soil_moisture < threshold:

   Trigger_Autonomous_Irrigation()

   if crop_health_status == "Pest Detected":

   Trigger_Autonomous_Pesticide_Spraying()

   if crop_health_status == "Ready to Harvest":

   Activate_Robotic_Harvester()

8.3.3 AAR implementation

8.3.3.1 Independent path planning and navigation

Autonomous navigation algorithms are used to make it possible for drones and robotic vehicles to work effectively in agricultural fields. These consist of: path planning using GPS for accurate row navigation; obstacle detection using LiDAR to prevent collisions; 3D field maps produced using Simultaneous Localization and Mapping (SLAM). Sensor fusion technology is used in robots and drones to increase mobility and operation accuracy.

8.3.3.2 *AI and IoT sensors for automated crop monitoring*

High-resolution photos of crops taken by drones are subjected to deep learning analysis in order to identify water stress, pest infestations, and nutrient deficits. For AI-driven analytics, IoT sensors send real-time temperature, humidity, and soil moisture data to cloud servers. Farmers can take prompt action when crucial crop health risks are brought to their attention by automated warnings.

8.3.3.3 *Intelligent water management and accurate irrigation*

Up to 40% less water is wasted when irrigation programs are automated using soil moisture data. Crops are given the appropriate amount of water at the appropriate time thanks to AI-driven decision-making. Wireless actuators and IoT nodes are used to operate drip irrigation systems.

8.3.4 Automated farming operations and robotic harvesting

Robotic arms driven by AI determine when fruit is ripe and carry out automated harvesting. Precision weeding, made possible by machine vision algorithms, reduces the need for herbicides by 60%. Autonomous tractors carry out fertilizing, sowing, and plowing.

```
Input: Pseudocode for Autonomous Navigation and Path Planning
  1. Autonomous Navigation for Agricultural Robot
     Initialize_GPS()
     Initialize_LiDAR()
     while Robot_Active:
     current_position = GPS_Get_Coordinates()
     obstacle_detected = LiDAR_Scan_Area()
     if obstacle_detected:
     Avoid_Obstacle()
  2. Follow Pre-defined Route Based on GPS Data
     next_waypoint = Get_Next_Waypoint()
     Move_To_Waypoint(next_waypoint)
     if Task_Completed():
     Stop_Robot()
```

Table 8.1 Functions used in the AI models for better accuracy

Step	*Description*	*Parameters*
1	Retrieve historical and current yield data	total_yield_before = Get_Historical_Yield() total_yield_after = Measure_Crop_Yield()
2	Calculate crop yield improvement	yield_improvement = ((total_yield_after - total_yield_before) / total_yield_before) * 100
3	Measure water and pesticide consumption using IoT sensors	water_usage = Measure_Water_Consumption() pesticide_usage = Measure_Pesticide_Consumption()
4	Compare measured values with traditional farming methods	water_savings = Compare_To_Traditional("Water", water_usage) pesticide_savings = Compare_To_Traditional("Pesticides", pesticide_usage)
5	Output performance metrics	Print("Yield Improvement:", yield_improvement, "%") Print("Water Savings:", water_savings, "%") Print("Pesticide Reduction:", pesticide_savings, "%")

8.4 EXPERIMENTAL ANALYSIS AND PERFORMANCE EVALUATION

A number of performance criteria are assessed in order to verify the efficacy of IoT-based precision farming and AAR. Calculate the percentage increase in crop output brought about by precision farming practices.

Efficiency of Resources: Determine whether water, fertilizer, and pesticide use have decreased.

Reduction of Operational Costs: Examine labor, fuel, and manual farm operations cost savings.

Accuracy of AI Models: Analyze the accuracy of AI-based models for yield prediction, crop classification, and disease detection.

Accuracy of Autonomous Navigation: Calculate the error rate in robotic obstacle avoidance and path planning using GPS.

Usability and Adoption by Farmers: To evaluate the usability and farmer acceptance of IoT-driven robotics, conduct surveys and analyze feedback (Table 8.1).

8.5 DIFFICULTIES AND RESTRICTIONS

Notwithstanding the encouraging developments, a number of issues still need to be resolved: It takes a large investment in hardware and AI development to implement autonomous robotics. Remote agricultural areas might not have high-speed internet, which is necessary for IoT-based devices. Farming systems powered by IoT are susceptible to data leaks and hackers, which lightweight cryptographic and privacy-preserving techniques can help mitigate (Bandaru et al., 2024b). The technical know-how needed to control AI-powered agricultural robots is often lacking among farmers. Future studies should concentrate on improving cybersecurity, coming up with affordable solutions, and designing intuitive robotic interfaces to close the gap between AI technology and practical farming applications (Table 8.2).

The suggested technique offers a methodical way to develop, deploy, and assess IoT-driven smart farming solutions including AAR. The project intends to improve modern agriculture's sustainability, maximize resource use, and boost productivity by utilizing AI, machine vision, IoT sensors, and real-time analytics. Next-generation precision agricultural solutions will be made possible by the experimental validation, which will evaluate effectiveness, accuracy, and practical applicability.

Table 8.2 Comprehensive methodology framework for autonomous agricultural robotics and IoT in precision farming

Parameter	*Description*	*Techniques used*	*Expected outcome*
Robotic Components	Autonomous robots for irrigation, spraying, weeding, and harvesting	GPS-based navigation, LiDAR, Machine Vision	Reduced labor dependency, high precision farming
IoT Sensor Network	Soil moisture, temperature, humidity, and multispectral imaging sensors for real-time data collection	Wireless sensor networks (WSNs), Edge Computing	Real-time monitoring, efficient resource usage
AI Model for Decision-Making	AI-driven predictive models for disease detection, irrigation scheduling, and crop health monitoring	Machine Learning (CNN, RNN), Predictive Analytics	Proactive interventions, reduced crop losses
Cloud-Based Data Processing	Centralized storage and analysis of farm data for real-time decision-making	AWS IoT, Google Cloud, Microsoft Azure FarmBeats	Scalable data management, automated analytics
Autonomous Navigation	Self-driving robotic tractors and drones for field monitoring and automated farming operations	GPS, LiDAR, SLAM (Simultaneous Localization and Mapping)	Improved field efficiency, reduced operational time
Automated Crop Monitoring	AI-based analysis of plant health using multispectral imaging and deep learning models	CNN (Convolutional Neural Networks), Image Processing	Early disease detection, optimized plant health

(Continued)

Table 8.2 (Continued) Comprehensive methodology framework for autonomous agricultural robotics and IoT in precision farming

Parameter	*Description*	*Techniques used*	*Expected outcome*
Precision Irrigation	Water conservation through soil moisture sensing and automated irrigation	IoT-Based Drip Irrigation, AI-Powered Decision Support	Water savings, reduced over-irrigation
Pest and Disease Management	Smart pesticide spraying using drones based on real-time pest detection	UAV-based remote sensing, AI-Powered Pest Recognition	Reduced pesticide usage, improved crop protection
Automated Harvesting	Robot-assisted harvesting using AI-powered robotic arms	Machine Vision, Deep Learning-Based Ripeness Detection	Reduced harvesting costs, optimized labor
Crop Yield Improvement	Increase in productivity due to AI-driven precision farming	Yield comparison before and after implementation	Higher crop production, improved quality
Resource Utilization	Measurement of water, pesticide, and fertilizer efficiency	IoT Sensors, Smart Actuators, Cloud Analytics	Reduced wastage, optimized usage
Cost Savings	Analysis of operational costs before and after automation	Cost-Benefit Analysis, ROI Estimation	Higher profit margins, affordability
AI Model Accuracy	Evaluation of AI-based predictive analytics for disease detection and irrigation	Performance Metrics (Precision, Recall, F1-Score)	Reliable AI-driven decision-making
Autonomous Navigation Efficiency	GPS-based path accuracy and obstacle detection performance	LiDAR Obstacle Avoidance, GPS-Based Route Mapping	Safe, efficient, and automated field operations
User Adoption and Feedback	Survey on farmer satisfaction and ease of use	User Experience Testing, Questionnaire Surveys	Improved user adoption, system enhancement

8.6 RESULTS AND DISCUSSION

The performance of the IoT-based Precision Farming System and AAR in comparison to conventional farming practices is assessed in this section. The experiment, which lasted 6 months, was carried out on a 100-hectare large-scale farm that produced wheat, corn, and tomatoes. The study concentrated on important agricultural metrics like increased crop output, disease diagnosis accuracy using AI models, efficient use of water and pesticides, and total cost savings (Table 8.3).

Table 8.3 Experiment setup for the proposed project

Parameter	*Description*
Field Size	100 Hectares (Large-Scale Farm)
Crops Cultivated	Wheat, Corn, Tomatoes
Experiment Duration	6 Months
Comparative Groups	Traditional Farming vs. IoT and Robotics-Based Precision Farming
Data Collection Tools	IoT Sensors, Drones, Cloud-Based Analytics, AI Models

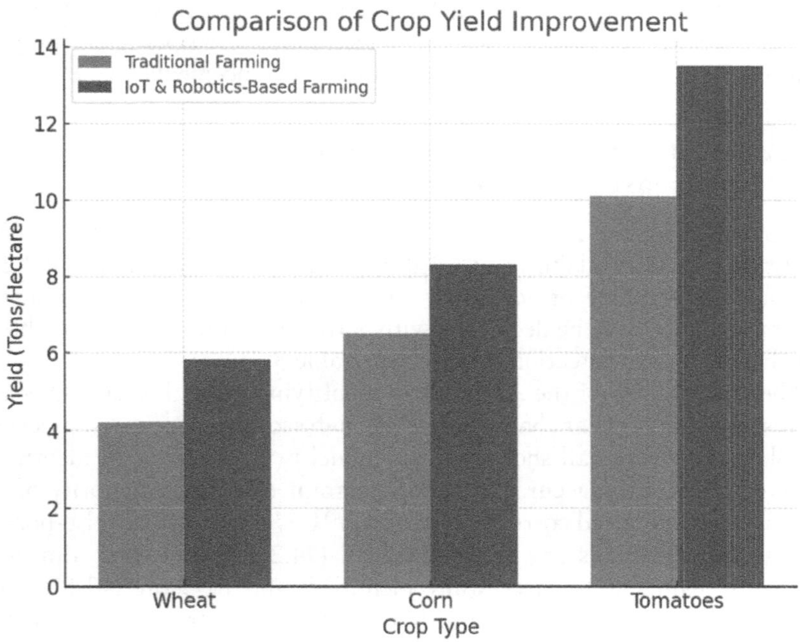

Figure 8.1 Comparison of crop yield improvement.

8.6.1 Results for crop yield improvement

According to the findings, implementing robotic automation and IoT-driven precision farming increased yield by an average of 33%. With a 38% production improvement, wheat outperformed the other crops under study, followed by corn (27%), and tomatoes (34%). The following elements are responsible for the increased yield. AI-based disease detection decreased crop loss by enabling early infection identification. Hydration was maximized using precision irrigation that delivered water according to the soil's current moisture content and reduced chemical stress on crops through targeted pesticide application, which ensured that only impacted plants were treated (Figure 8.1, Table 8.4).

Table 8.4 Crop yield improvement

Crop type	*Traditional yield (tons/hectare)*	*IoT and robotics-based yield (tons/hectare)*	*Yield improvement (%)*
Wheat	4.2	5.8	+38
Corn	6.5	8.3	+27
Tomatoes	10.1	13.5	+34

An average yield increase of 33% was achieved through the use of IoT-driven smart farming and autonomous robotics. Wheat saw the largest production increase (+38%), followed by corn (+27%) and tomatoes (+34%). Real-time AI-based disease detection, tailored pesticide spraying, and precision irrigation are credited with the advancements.

8.6.2 Results for AI model accuracy in disease detection

The accuracy of the AI-based disease detection system was 92.8%, demonstrating its dependability. According to a high recall score of 94.2%, the model is capable of accurately identifying the majority of unhealthy plants. Crop loss can be decreased with early intervention made possible by ML-based disease detection (Figure 8.2, Table 8.5).

The effectiveness of the AI model in identifying plant diseases is visually represented in the radar chart above. The high scores for accuracy, precision, recall, and F1-Score all show that the model is very reliable in identifying infected crops early; accuracy (92.8%) guarantees that the majority of disease cases are detected correctly; precision (91.5%) means that false-positive disease identifications are minimal; recall (94.2%) shows that almost all unhealthy plants were successfully identified; and F1-Score (92.8%) confirms a balanced performance across all evaluation metrics. With such high performance, AI-based disease detection minimizes crop loss, maximizes the use of pesticides, and promotes sustainable farming practices.

8.6.3 Discussion on benefits of robotic precision farming

The line chart illustrates the key improvements in crop yield, water usage, pesticide application, and labor costs due to IoT and robotics-based farming (Table 8.6). Crop yield increased from 6.93 to 9.2 tons per hectare (+33%) due to real-time AI monitoring and automation. Water usage dropped by 34.6%, thanks to smart irrigation technologies that optimize water distribution. Pesticide use reduced by 42%, as AI-powered pest detection ensured targeted pesticide spraying. Due to the replacement of manual farm labor by autonomous robots, labor costs decreased by 40% (Figure 8.3).

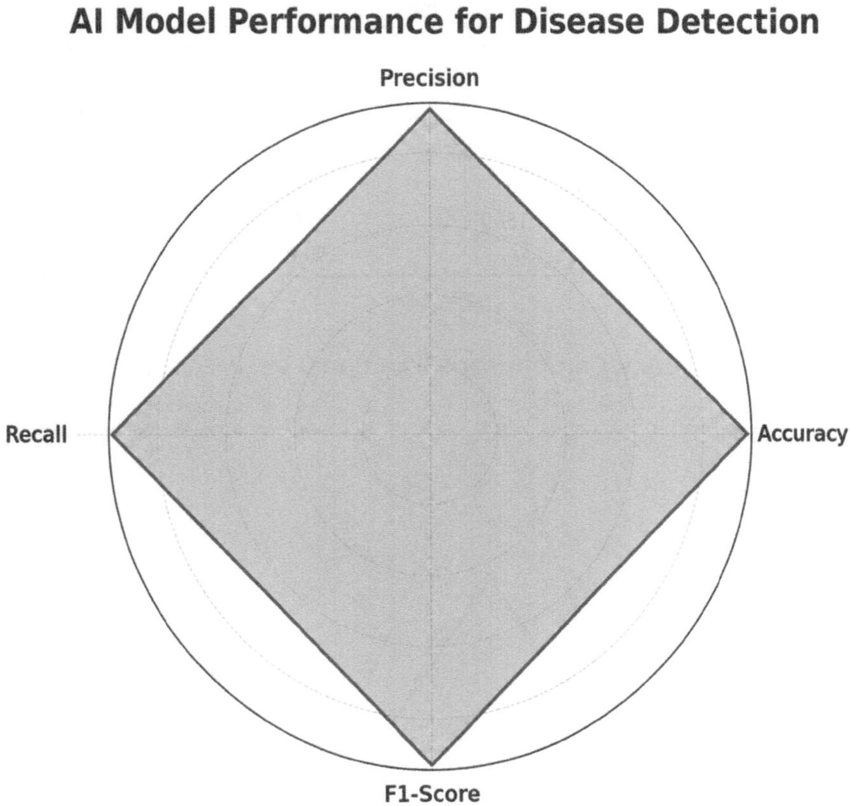

Figure 8.2 AI model performance results.

Table 8.5 AI model performance for crop disease detection

Performance metric	*Value (%)*
Accuracy	92.8
Precision	91.5
Recall	94.2
F1-Score	92.8

8.7 CONCLUSION AND FUTURE WORK

Significant gains in productivity, efficiency, and sustainability have been shown when AAR and IoT-based Precision Farming are combined. According to the study's findings, crop productivity increased by 33%, water consumption decreased by 34.6%, pesticide use decreased by 42%, and labor

Table 8.6 Benefits of robotic precision farming

Parameter	*Traditional farming*	*IoT and robotics-based farming*	*Improvement (%)*
Crop Yield (Tons/Hectare)	6.93 (average)	9.2 (average)	+33
Water Usage (Liters/Hectare)	120,000	78,500	−34.6
Pesticide Use (Liters/Hectare)	25	14.5	−42
Labor Cost (USD/Hectare)	1500	900	−40
AI Model Accuracy (%)	--	92.8	–

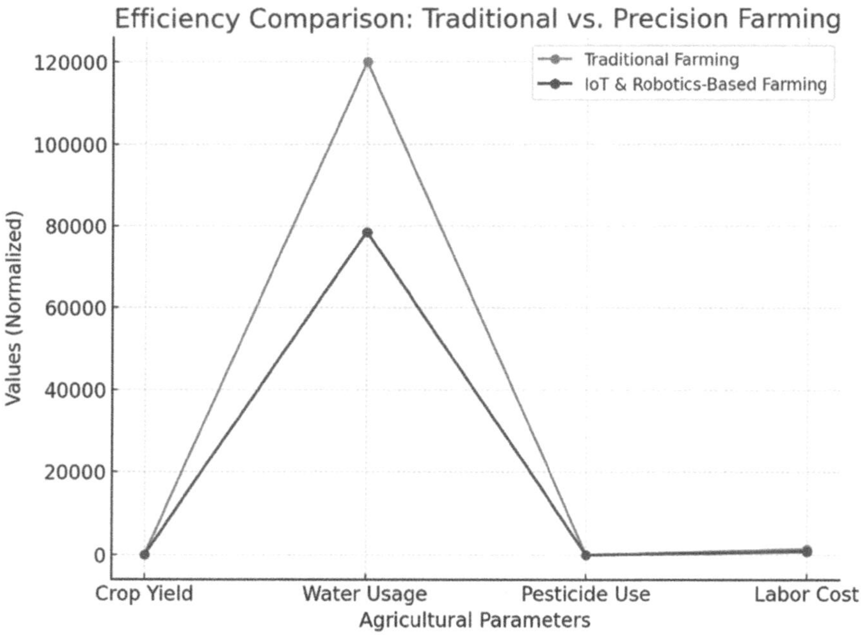

Figure 8.3 Efficiency comparison.

costs decreased by 40%. With a 92.8% accuracy rate, the AI-driven disease detection model reduced crop losses and allowed for early intervention. Real-time AI monitoring, targeted pesticide spraying, and precision irrigation have all improved resource use, making farming more economical and ecologically benign. The advantages of autonomous farming technology exceed its drawbacks, even with its high initial cost and communication issues. Agriculture could undergo a revolution with the use of robotic automation and AI-driven analytics, guaranteeing environmental preservation, economic viability, and food security. According to the study's findings, smart farming based on IoT and robotics will be essential in determining how precision agriculture develops in the future and will make it more sustainable, scalable, and efficient.

Future studies should concentrate on blockchain-based data security for agricultural analytics (Bandaru & Visalakshi, 2023), integrating renewable energy for sustainable automation, and developing affordable AI-driven robotics solutions for small-scale farmers. Swarm robotics, edge computing, and predictive analytics developments can all improve autonomous decision-making, increasing the accessibility and global scalability of smart farming.

ACKNOWLEDGMENTS

We are deeply grateful to Dr. M. Venu, Professor and Principal, Mrs. M. Sri Lakshmi, Vice Principal, and Dr. Prakash Pareek for their essential help, encouragement, and direction during this project. Additionally, we would like to thank Vishnu Institute of Technology, Bhimavaram, for providing the environment and resources required to complete this job. Lastly, we express our gratitude to our families, classmates, and mentors for their unwavering encouragement and support.

REFERENCES

Albiero, D., Garcia, A. P., Umezu, C. K., & de Paulo, R. L. (2022). Swarm robots in mechanized agricultural operations: A review about challenges for research. *Computers and Electronics in Agriculture*, 193, 106608. https://doi.org/10.1016/j.compag.2021.106608

Arad, A., Ben-Shahar, O., & Shoham, M. (2021). Deep-learning-based robotic harvesting for high-throughput agriculture. *Computers and Electronics in Agriculture*, 186, 106136. https://doi.org/10.1016/j.compag.2021.106136

Ayaz, M., Ammad-Uddin, M., Sharif, Z., Mansour, A., & Aggoune, E.-H. M. (2019). Internet-of-Things (IoT)-based smart agriculture: Toward making the fields talk. *IEEE Access*, 7, 129551–129583. https://doi.org/10.1109/ACCESS.2019.2932609

Bac, C. W., Hemming, J., & Edan, Y. (2014). Harvesting robots for high-value crops: State-of-the-art review and challenges ahead. *Journal of Field Robotics*, 31(6), 888–911. https://doi.org/10.1002/rob.21544

Bandaru, V. N. R., & Visalakshi, P. (2023, June). BDCT-Blockchain-Based Decentralized Computing and Tamper Resistance for Cloud Storage. In *2023 International Conference on Advanced & Global Engineering Challenges (AGEC)* (pp. 71–77). Surampalem, India: IEEE.

Bandaru, V. N. R., Kaligotla, V. G. S., Varma, U. D. S. P., Prasadaraju, K., & Sugumaran, S. (2024a, July). A Enhancing Data Security Solutions for Smart Energy Systems in IoT-Enabled Cloud Computing Environments through Lightweight Cryptographic Techniques. In *IOP Conference Series: Earth and Environmental Science* (Vol. 1375, No. 1, p. 012003). IOP Publishing. https://doi.org/10.1088/1755-1315/1375/1/012003

Bandaru, V. N. R., Sumalatha, M., Rafee, S. M., Prasadraju, K., & Sri, M. (2024b). Enhancing privacy measures in healthcare within cyber-physical systems through cryptographic solutions. *EAI Endorsed Transactions on Scalable Information Systems*, 11(6), Article 5732. https://doi.org/10.4108/eetsis.5732.

Bechar, A., & Vigneault, C. (2016). Agricultural robots for field operations: Concepts and components. *Biosystems Engineering*, 149, 94–111. https://doi.org/10.1016/j.biosystemseng.2016.06.014

Duckett, T., Pearson, S., Blackmore, S., & Grieve, B. (2018). Agricultural robotics: The future of robotic agriculture. UK-RAS Network White Papers. arXiv:1806.06762. https://arxiv.org/abs/1806.06762

Farooq, M. S., Riaz, S., Abid, A., Umer, T., & Zikria, Y. B. (2020). Role of IoT technology in agriculture: A systematic literature review. *Electronics*, 9(2), 319. https://doi.org/10.3390/electronics9020319

Jain, R., Kumar, P., & Singh, R. (2021). Role of 5G in agricultural IoT applications: A systematic review. *Internet of Things*, 14, 100390. https://doi.org/10.1016/j.iot.2021.100390

Ju, C., Kim, J., Seol, J., & Son, H. I. (2022). A review on multi-robot systems in agriculture. *Computers and Electronics in Agriculture*, 202, 107336. https://doi.org/10.1016/j.compag.2022.107336

Kamilaris, A., & Prenafeta-Boldú, F. X. (2018). Deep learning in agriculture: A survey. *Computers and Electronics in Agriculture*, 147, 70–90. https://doi.org/10.1016/j.compag.2018.02.016

Kumar, N., Pande, S., & Rana, P. S. (2020). Cloud computing in precision agriculture: Challenges and solutions. *Computers and Electronics in Agriculture*, 169, 105238. https://doi.org/10.1016/j.compag.2020.105238

Liakos, K. G., Busato, P., Moshou, D., Pearson, S., & Bochtis, D. (2018). Machine learning in agriculture: A review. *Sensors*, 18(8), 2674. https://doi.org/10.3390/s18082674

Mandal, R., Patel, K., & Saha, S. (2018). AI-based precision irrigation and its impact on water conservation. *Environmental Science & Policy*, 86, 74–84. https://doi.org/10.1016/j.envsci.2018.05.004

Mohanty, S. P., Hughes, D. P., & Salathé, M. (2016). Using deep learning for image-based plant disease detection. *Frontiers in Plant Science*, 7, 1419. https://doi.org/10.3389/fpls.2016.01419

Patil, S., Tiwari, A., & Sharma, M. (2016). IoT-based smart irrigation system for sustainable agriculture. *Procedia Computer Science*, 92, 155–160. https://doi.org/10.1016/j.procs.2016.07.160

Chapter 9

IoT and Edge AI-enabled next-generation agriculture

Rardchawadee Silapunt

9.1 INTERNET OF THINGS IN THE REALM OF AGRICULTURE

Bringing the Internet of Things (IoT) into agriculture is driving a major transformation in farming methods, improving efficiency, productivity, and sustainability. IoT refers to a web of connected devices equipped with sensors, software, and related technologies that gather and share data through the internet. In agriculture, such devices track soil moisture, crop health, weather, and livestock conditions, helping farmers make informed decisions and act promptly when needed. Figure 9.1 shows practical IoT applications in agriculture followed by details of each example.

1. Precision Farming and Input Management: IoT technologies play a central role in precision farming by improving field-level management of crops. With the help of sensors and data analytics, farmers can observe crop conditions in real time and apply water, fertilizer,

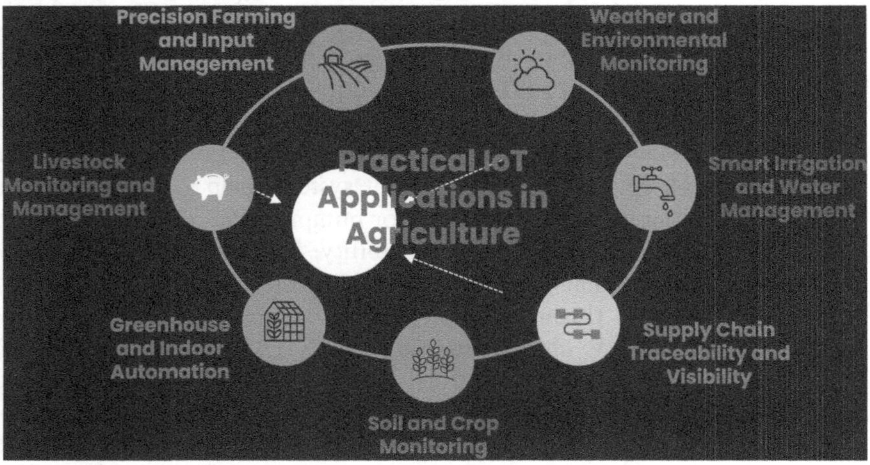

Figure 9.1 Practical IoT applications in agriculture.

DOI: 10.1201/9781003589730-9

and pesticides with greater accuracy. This site-specific strategy cuts down on waste while increasing both yield and quality. In addition, IoT-driven precision agriculture promotes more efficient resource use and lowers environmental impact.

2. Weather and Environmental Monitoring: Sensors deployed in agricultural fields gather data on environmental conditions such as temperature, humidity, rainfall, wind speed, and solar radiation. This information helps farmers understand the microclimate of their fields and make data-driven decisions regarding planting, harvesting, and crop protection. Monitoring environmental conditions plays an important role in anticipating and reducing weather-related risks, including frost damage, heat stress, and disease pressure [1]. Long-term environmental monitoring also contributes to climate adaptation strategies. By analyzing trends in temperature, precipitation, and growing season length, farmers can adjust crop selections, planting dates, and management practices to adapt to changing climate conditions. Furthermore, continuous monitoring of greenhouse gas emissions and soil carbon sequestration supports sustainable farming practices aligned with climate mitigation goals [1,2].
3. Smart Irrigation and Water Management: In agriculture, the challenge of water scarcity is becoming increasingly serious, highlighting the need for effective water management to achieve sustainable production. IoT-enabled smart irrigation systems help optimize water use by tracking soil moisture, weather patterns, and crop requirements. Using real-time data, these systems automatically regulate irrigation timing and water distribution so that crops receive the proper amount when needed. This approach conserves substantial amounts of water while sustaining or even enhancing crop health and yields. Using IoT-enabled irrigation can reduce water consumption by as much as 30% while simultaneously boosting crop productivity. These systems consider multiple factors including soil moisture content, evapotranspiration rates, weather forecasts, and crop-specific water needs to deliver precisely calibrated irrigation. By avoiding both over-irrigation and under-irrigation, such intelligent systems help save water while also lowering the energy required for pumping and distribution [3].
4. Supply Chain Traceability and Visibility: IoT enhances agricultural supply chain management by improving traceability, logistics, and quality control throughout the journey from farm to consumer. Radio Frequency Identification (RFID) tags, Global Positioning System (GPS) tracking, and environmental sensors monitor the condition of agricultural products during transportation and storage, ensuring food safety and reducing post-harvest losses. This increased visibility throughout the supply chain enables better inventory management, reduced waste, and improved response to market

demands. The integration of IoT in agricultural supply chains creates transparency that benefits all stakeholders. Producers can verify that their products are handled properly, processors receive consistent quality inputs, retailers can track product freshness, and consumers gain access to information about product origins and handling. This transparency builds trust and supports premium pricing for quality-assured products [3].

5. Soil and Crop Monitoring: The application of IoT in precision farming extends to soil analysis, crop monitoring, and yield prediction. Sensors deployed in fields collect data on soil properties such as moisture, temperature, and nutrient content, allowing farmers to make data-driven decisions regarding planting, fertilization, and irrigation. This granular approach to farm management represents a significant advancement over traditional farming methods that rely on uniform applications across entire fields [4].
6. Greenhouse and Indoor Automation: IoT applications in greenhouse and indoor agriculture enable automated, real-time monitoring and control of key environmental factors such as temperature, humidity, soil moisture, light intensity, and CO_2 levels. By deploying interconnected sensors and actuators, these systems continuously collect data and automatically adjust climate control, irrigation, and lighting to maintain optimal growing conditions. This automation not only improves crop yield and quality but also enhances resource efficiency by ensuring water, energy, and nutrients are used only as needed, significantly reducing waste and operational costs. Furthermore, IoT-based greenhouse systems offer remote management capabilities through cloud platforms and mobile applications, allowing farmers to monitor conditions, receive alerts, and adjust settings from anywhere.
7. Livestock Monitoring and Management: IoT applications in livestock management have transformed how farmers monitor and care for their animals. Wearable IoT devices track animal health, movement, and behavior, providing real-time insights into individual animals and herds. These devices can monitor vital signs, detect early signs of disease, track feeding patterns, and even predict reproductive cycles. The continuous monitoring capabilities enable early intervention for health issues, optimized feeding schedules, and improved breeding management [5]. Livestock monitoring systems also enable location tracking, which is particularly valuable for free-range animals. RFID tags and GPS-enabled collars allow farmers to monitor animal movements, detect straying behavior, and manage grazing patterns. Additionally, automated feeding systems can dispense customized nutrition based on individual animal needs, optimizing feed efficiency and promoting animal health [4].

9.2 ARTIFICIAL INTELLIGENCE IN AGRICULTURE

Advances in technology are driving a major shift within the agricultural sector, especially the Artificial Intelligence (AI) and the emerging trend of Edge AI. As global population growth intensifies the demand for food production, AI technologies are being increasingly adopted to enhance efficiency, sustainability, and productivity in farming practices. AI's applications in agriculture are diverse and impactful, encompassing several key areas.

1. Crop Health Monitoring: AI systems analyze vast amounts of data from sensors, satellite imagery, and historical records to optimize farming practices. This includes managing soil health, determining optimal planting times, and applying fertilizers and pesticides more efficiently. For instance, AI algorithms can process real-time data to adjust irrigation schedules based on soil moisture levels and weather forecasts, thereby conserving water while ensuring crops receive adequate hydration [6]. AI-powered image recognition systems can identify plant diseases, nutrient deficiencies, and pest infestations at early stages, often before they become visible to the human eye. These systems analyze leaf coloration, growth patterns, and subtle visual cues that indicate plant stress or disease, allowing for targeted interventions that minimize crop damage and reduce the need for broadcast applications of pesticides. Machine learning algorithms continuously improve their diagnostic accuracy as they process more data, creating increasingly sophisticated crop monitoring capabilities [7].
2. Predictive Analytics: AI models utilize historical data and machine learning techniques to forecast weather patterns, crop yields, and market trends. This predictive capability enables farmers to make informed decisions regarding planting schedules, resource allocation, and pest management. By anticipating challenges such as pest outbreaks or adverse weather conditions, farmers can implement proactive measures to safeguard their crops [4]. Predictive analytics extends to yield forecasting, which supports both on-farm planning and broader market dynamics. By analyzing historical yield data alongside current growing conditions, AI can provide increasingly accurate predictions of harvest volumes and timing. These predictions support better labor planning, equipment scheduling, and marketing strategies. At a broader scale, aggregate yield predictions inform supply chain planning, price forecasting, and food security assessments [1,4].
3. Autonomous Farming Equipment: The integration of AI-powered machinery automates various agricultural tasks, from planting to harvesting. Robotic systems equipped with computer vision can identify ripe produce and harvest it with precision, significantly reducing labor costs and minimizing crop damage during the harvesting process. These autonomous systems operate with consistency and precision

that often exceeds human capabilities, particularly for repetitive tasks that require sustained attention. Autonomous tractors and specialized agricultural robots represent a frontier in farm automation. These machines utilize GPS guidance, obstacle detection sensors, and sophisticated control systems to perform field operations with minimal human intervention. They can follow optimized paths through fields, adjust application rates in real time based on soil and crop conditions, and operate continuously during critical weather windows. The advancement of these technologies addresses both labor shortages in agricultural regions and the increasing precision requirements of modern farming [8].

4. Market Analysis: AI helps farmers and agribusinesses analyze market trends, predict demand, and optimize pricing strategies. By processing vast amounts of data from diverse sources—including historical price records, weather patterns, global production volumes, and consumer trends—AI systems provide insights that support strategic decision-making. These insights enable farmers to identify optimal times to sell their products, select crops with favorable market outlooks, and adjust production practices to meet specific market requirements. Market analysis also extends to consumer preference modeling, which is increasingly important as agricultural markets fragment into specialized segments. AI algorithms can identify emerging consumer trends, predict preference shifts, and quantify willingness-to-pay for specific product attributes. This information helps farmers align production practices with market demands, potentially capturing premium prices for attributes such as organic certification, sustainability credentials, or unique quality characteristics [3].

9.3 THE RISE OF EDGE AI IN AGRICULTURE

As farming continues its shift toward digital transformation, Edge AI is becoming a vital element of this change. In this context, Edge AI means running AI models directly on devices located near the point of data collection, instead of depending only on centralized cloud systems. This method provides several important benefits that are especially relevant in agriculture. Figure 9.2 illustrates an IoT framework where Edge AI devices link to a cloud server through a Low-Power Wide Area Network (LoRaWAN). Below are the key benefits of Edge AI.

1. Real-Time Processing: With Edge AI, data can be processed directly on-site, enabling farmers to act instantly on changing conditions without delays from cloud-based systems. For example, drones using Edge AI are able to assess crop health in real time and automatically deliver treatments when required. By applying these systems,

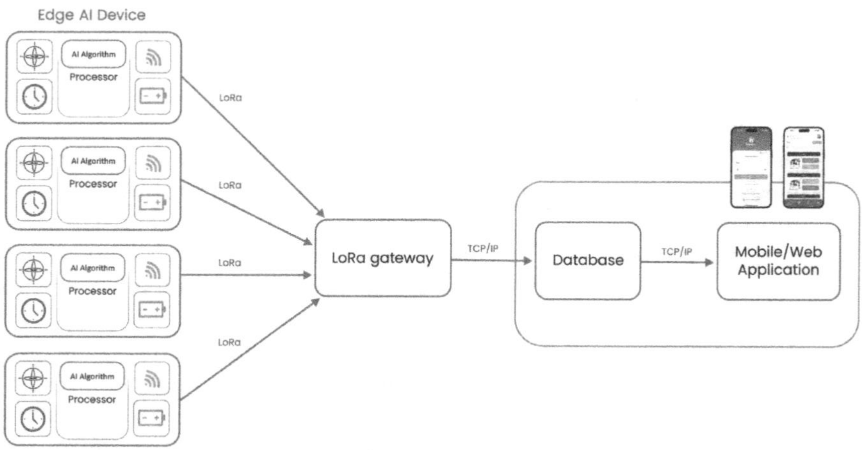

Figure 9.2 An example of IoT architecture consisting of Edge AI devices.

irrigation systems can make instantaneous adjustments based on soil moisture readings, and livestock monitoring systems can immediately alert farmers to health issues or abnormal behaviors [9]. The ability to process data in real time is crucial for time-sensitive agricultural operations. During harvest, Edge AI-equipped machinery can assess crop ripeness and quality as it operates, adjusting settings to optimize harvest quality and minimize losses. During planting, similar systems can monitor seed placement and soil conditions, making immediate adjustments to ensure optimal germination conditions. These capabilities transform reactive farming practices into proactive, optimized operations that maximize efficiency and productivity [9].

2. Reduced Bandwidth Requirements: By processing data locally, Edge AI reduces the need for constant internet connectivity and minimizes bandwidth usage. This is particularly advantageous in remote agricultural areas where internet access may be limited or unreliable. Instead of sending all raw data to cloud servers, Edge AI units analyze the information locally and forward only essential outcomes or alerts, which greatly cuts down the amount of data transferred and the bandwidth needed [9]. The reduced dependence on continuous connectivity makes Edge AI solutions more robust and reliable in agricultural settings. Farm operations can continue uninterrupted even during network outages, and systems remain functional in areas with spotty coverage. This reliability is essential for critical applications such as irrigation control, livestock monitoring, and greenhouse management, where system failures could result in significant losses [2,9].

3. Enhanced Privacy and Security: Local data processing enhances data privacy by minimizing the transmission of sensitive information over the internet. This is crucial for protecting proprietary farming techniques and sensitive operational data. Edge AI systems can analyze and act upon data without exposing it to potential security vulnerabilities associated with cloud transmission and storage [10]. Privacy considerations are increasingly important as agricultural data becomes more valuable. Farm operational data can reveal proprietary practices, business strategies, and competitive advantages that farmers wish to protect. By processing sensitive data locally and transmitting only necessary information, Edge AI systems provide a layer of privacy protection that centralized cloud systems cannot match. This approach also simplifies compliance with data protection regulations that restrict the transmission of certain types of information across jurisdictional boundaries [10].

9.4 INTEGRATION OF EDGE AI WITH IoT IN AGRICULTURE

Combining Edge AI with IoT is transforming modern farming by bringing new levels of precision, efficiency, and sustainability. With rising global food demand and mounting climate pressures, this synergy enables farmers to manage resources more effectively, boost yields, and lessen environmental impacts through real-time, data-driven insights. Because Edge AI processes information directly on IoT devices instead of depending entirely on cloud servers, it reduces delays, strengthens data privacy, and supports autonomous functions—even in remote areas where connectivity is limited.

IoT networks in agriculture consist of interconnected sensors, drones, and machinery that collect vast amounts of agricultural data—soil moisture, temperature, crop health, livestock behavior, and weather patterns. Edge AI processes this data directly on the devices or nearby edge servers, eliminating the need for constant cloud connectivity. For example, a soil sensor with embedded AI can analyze moisture levels locally and trigger irrigation systems without waiting for remote server instructions [2,11]. This decentralized approach ensures rapid responses to field conditions while reducing bandwidth costs and data vulnerabilities.

9.5 TECHNICAL CHALLENGES IN BUILDING EDGE AI SYSTEMS FOR AGRICULTURE

The development and deployment of Edge AI systems in agriculture involve navigating a maze of technical challenges, from hardware constraints to algorithmic adaptability. Below is a detailed breakdown of the key complexities.

1. Hardware Limitations in Harsh Environments: Edge AI devices in agriculture must operate reliably in extreme conditions (heat, moisture, dust) while maintaining computational efficiency. Ruggedized hardware, e.g., NVIDIA Jetson AGX Orin is required to withstand environmental stressors, but this increases costs. Passive cooling systems are often inadequate for sustained AI workloads, leading to thermal throttling. Balancing performance with energy constraints is also challenging. For example, running real-time object detection (YOLOv8) on a drone demands high compute power, which strains battery life.
2. AI Model Optimization for Resource-Constrained Devices: Deploying AI models on edge devices requires drastic optimization to fit limited memory and processing capabilities. Reducing model precision (32-bit → 8-bit) cuts memory usage by 75% but risks accuracy loss. Pruning redundant neurons shrinks models but requires retraining. Apart from that, lightweight models like MobileNet or TinyML trade versatility for efficiency, struggling with complex tasks, e.g., multispectral crop analysis. In addition, implementing novel federated learning for localized model updates is hampered by limited compute resources and intermittent connectivity.
3. Data Heterogeneity and Quality: Agricultural data varies widely across crops, soils, and climates, complicating model generalization. Soil moisture sensors may malfunction in clay-rich soils, while camera-based systems fail in low-light conditions. Labeling datasets is usually labor-intensive and inconsistent annotations reduce model reliability. Differences in climate conditions also affect the trained model. For example, a pest detection model trained in temperate regions may fail in tropical climates without recalibration.
4. Connectivity and Interoperability: Remote rural areas may lack reliable internet connectivity, limiting the effectiveness of IoT applications that require constant data transmission. While Edge AI mitigates some connectivity challenges through local processing, many agricultural IoT systems still rely on periodic data transmission for updates, aggregated analytics, and remote monitoring. The digital divide between urban and rural areas creates inequitable access to technological benefits. Connectivity and interoperability challenges vary by region and technology type. In some areas, basic cellular coverage may be lacking, while in others, bandwidth limitations may restrict data-intensive applications. Solutions include the development of low-bandwidth protocols optimized for agricultural applications, deployment of alternative connectivity technologies such as LoRaWAN, NB-IoT or satellite communications, investment in rural broadband infrastructure,

and design of systems that gracefully handle intermittent connectivity without compromising core functionality [12].

5. Security and Privacy Risks: Sensitive farm data processed locally remains vulnerable to cyberattacks and physical tampering. ARM TrustZone, which is a hardware-based security extension that divides an ARM processor's resources into a "Secure World" and a "Non-Secure World", isolates sensitive operations but increases hardware costs. Protecting model IP via homomorphic encryption adds computational overhead, conflicting with latency requirements.
6. Power Management for Remote Deployments: Sustaining Edge AI devices in off-grid areas demands innovative power strategies. For instance, solar panels struggle in cloudy regions, while kinetic energy harvesters are unreliable for high-power devices. Periodically deactivating non-critical sensors, e.g., nighttime soil probes, conserves energy but risks missing data.
7. Scalability and Maintenance: Deploying thousands of edge devices across large farms introduces logistical hurdles. Rolling out firmware patches to devices with sporadic connectivity risks version fragmentation. Sensor failure detection requires embedding diagnostic AI, increasing model complexity.

9.6 ECONOMIC CHALLENGES

For small-scale farmers, the expense of IoT devices, AI applications, and the required infrastructure can be overwhelming. Upfront spending goes beyond hardware and software to include setup, customization, and integration with existing farming operations. In many developing areas, such high costs remain a major obstacle to adoption, even though the technology offers substantial long-term advantages.

The investment challenge is compounded by the perceived risk of new technologies. Farmers may be hesitant to invest in solutions without clear evidence of returns in their specific context. Additionally, the rapid pace of technological advancement creates concerns about obsolescence, where today's investments might be outpaced by tomorrow's innovations. Addressing these concerns requires financing mechanisms, cost-sharing arrangements, and demonstration projects that reduce financial risks for early adopters.

9.7 DATA PRIVACY AND SECURITY CHALLENGES

Ensuring the security of agricultural data and preventing cyberthreats is a growing concern as farm operations become increasingly digitized. IoT networks create potential vulnerabilities that could expose sensitive farm data or even compromise control systems. Farmers are rightfully concerned about who has access to their operational data and how it might be used by equipment manufacturers, service providers, or other entities.

Security challenges include protecting data transmission between sensors and processing systems, securing stored data against unauthorized access, and safeguarding control systems against malicious interference. Privacy concerns include clarity about data ownership, transparency regarding data usage by service providers, and control over how farm data might be aggregated or shared with third parties. Addressing these concerns requires robust technical security measures, clear data ownership agreements, and regulatory frameworks that protect farmer interests.

9.8 TECHNICAL EXPERTISE CHALLENGES

Farmers may require significant training and technical knowledge to effectively use IoT and AI technologies. The complexity of modern agricultural technology systems demands skills that traditional agricultural education may not provide. This knowledge gap can hinder adoption and effective utilization of available technologies, limiting their potential benefits.

The expertise challenge encompasses multiple dimensions. Installation and configuration of IoT systems require technical skills. Interpretation of AI-generated insights requires analytical capabilities. Troubleshooting and maintenance demand diagnostic abilities. System integration requires understanding how different technologies interact. Addressing these knowledge requirements necessitates comprehensive education and training programs, simplified user interfaces, robust technical support services, and potentially new service models where technology providers manage complexity on behalf of farmers.

9.9 FUTURE PROSPECTS

Smart agriculture's future will be driven by continuous progress in IoT and AI, together with changing farming practices and market needs. The following emerging trends are expected to significantly reshape the agricultural sector in the years ahead.

1. 5G Connectivity: Faster and more reliable internet connectivity through 5G networks will enhance real-time data transmission and IoT performance in agricultural settings. The high-bandwidth, low-latency, and massive device connectivity capabilities of 5G will support more sophisticated applications, from high-resolution real-time video analytics to coordinated autonomous vehicle operations. This improved connectivity will be particularly transformative in rural areas that

have historically struggled with limited internet access. The impact of 5G extends beyond simple connectivity improvements to enable new categories of applications. High-bandwidth applications such as augmented reality for equipment maintenance or real-time video analysis of crop conditions become feasible. Low-latency capabilities support precise coordination between multiple autonomous vehicles working simultaneously in a field. Massive device connectivity accommodates dense sensor networks that provide unprecedented monitoring resolution. Together, these capabilities will unlock new possibilities in agricultural technology applications.

2. Blockchain Integration: Secure and transparent supply chains enabled by blockchain technology will improve food traceability and reduce fraud in agricultural markets. Blockchain creates immutable records of transactions and product movements, establishing trust among supply chain participants and consumers. This transparency can support premium pricing for verified attributes such as organic certification, sustainable production practices, or specific origins. Beyond basic traceability, blockchain enables sophisticated applications such as smart contracts that automatically execute payment when delivery conditions are met, tokenization of agricultural assets that facilitates investment and risk-sharing, and decentralized marketplaces that connect producers directly with consumers or processors. These applications potentially restructure agricultural value chains, redirecting more value to producers and creating new business models based on verified trust rather than intermediary relationships.
3. AI-Driven Decision Support Systems: More sophisticated AI models will offer better predictive analytics and autonomous decision-making capabilities for farm management. These systems will integrate multiple data sources, considering complex interactions between variables such as weather patterns, soil conditions, crop genetics, and market dynamics. The resulting recommendations will optimize across multiple objectives such as yield, quality, sustainability, and profitability. The evolution of agricultural decision support systems will likely follow a progression from advisory systems that provide recommendations for human implementation to semi-autonomous systems that implement routine decisions automatically while referring complex decisions to human operators, and eventually to fully autonomous systems that manage entire growing processes with minimal human intervention. This progression will be enabled by improvements in AI reliability, predictive accuracy, and the ability to explain recommendations in terms that build user trust.
4. Sustainable Solutions: Innovations in AI and IoT will increasingly focus on climate-smart agriculture and eco-friendly farming practices. These technologies will support precision resource management that minimizes environmental impacts while maintaining productivity.

Applications will include carbon sequestration monitoring, biodiversity management, ecosystem service valuation, and integrated pest management that minimizes chemical usage. Sustainability-focused applications will align with growing market demands for environmentally responsible production practices and regulatory frameworks that incentivize or mandate sustainable approaches. Technologies will help farmers document their environmental performance, potentially accessing premium markets or participating in ecosystem service payment programs. Additionally, AI and IoT will support adaptation to changing climate conditions through improved monitoring of environmental stressors and optimization of resilient farming practices.

9.10 CASE STUDIES AND KEY APPLICATIONS OF EDGE AI-IoT INTEGRATION

The theoretical benefits of IoT and Edge AI in agriculture are being realized through numerous real-world applications across diverse farming contexts. These case studies demonstrate the practical impact of these technologies on agricultural productivity, sustainability, and profitability.

1. Real-Time Analysis for Crop Health and Disease Detection: Cameras and drones equipped with Edge AI can monitor crops in real time, enabling the early detection of diseases, nutrient shortages, and pest problems. Across regions such as India and Europe, IoT sensors track soil moisture, temperature, and humidity, and Edge AI processes this information on-site to identify early indicators of fungal infection [13]. Farmers who have adopted these systems have seen a 30% reduction in disease incidence thanks to real-time alerts that enable targeted fungicide application. In Europe, drones equipped with Edge AI scan crops for disease, reducing fungicide use while maintaining yields. These systems use cameras and multispectral sensors processed by lightweight convolutional neural networks (CNNs) on devices like the NVIDIA Jetson AGX Orin, allowing for immediate, on-site analysis without relying on the cloud. In the EU's CrackSense project, IoT sensors combined with Edge AI predict fruit cracking risks in citrus and grapes, allowing preemptive interventions [14]. Figure 9.3 shows the mobile monitoring system of fruit cracking.
2. Precision Irrigation and Nutrient Management: Munir et al. [15] proposed a practical approach to water-efficient agriculture by integrating IoT sensor networks with edge computing. In the system as shown in Figure 9.4, environmental data such as soil moisture and temperature are collected by IoT sensors and processed locally on an edge server, rather than relying solely on cloud computing. The decision-making process is hybrid: 50% is based on an ontology-driven

Figure 9.3 Mobile fruit cracking monitoring system [14].

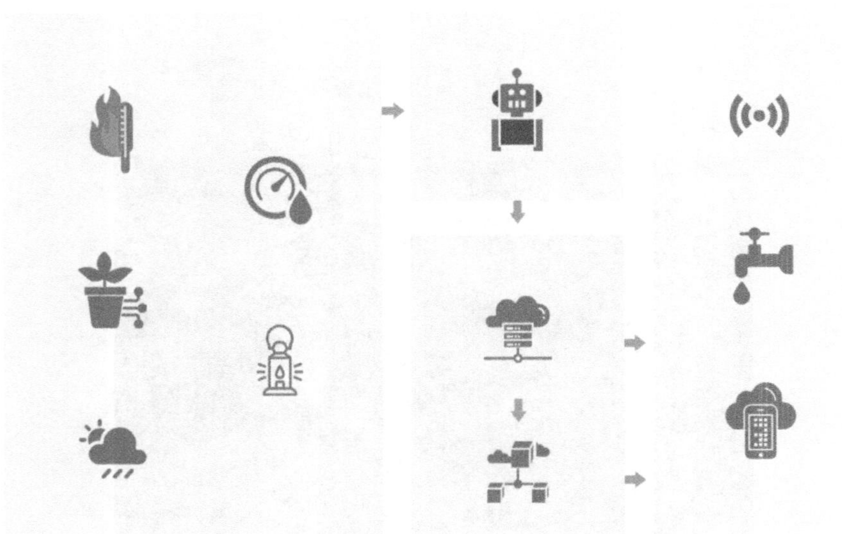

Figure 9.4 Proposed structure of the smart irrigation system [15].

rule system, and the other 50% uses real-time sensor data analyzed with a K-Nearest Neighbor (KNN) machine learning algorithm. This combination allows the system to optimize irrigation scheduling with minimal latency, reduce the processing burden on centralized servers, and function reliably even with intermittent connectivity.

3. Autonomous Farming Equipment: Neousys Technology has enabled the retrofitting of agricultural tractors with edge computers connected to cameras, sensors, and Light Detection and Ranging (LiDAR) to achieve autonomous operation [16]. These Edge AI-enabled tractors as shown in Figure 9.5 can be remotely monitored and controlled, allowing a single skilled operator to manage multiple vehicles simultaneously. The system processes data from the vehicle's sensors in real time, enabling precision navigation, obstacle detection, and dynamic adjustment of field operations without relying on cloud connectivity. This approach addresses labor shortages, increases operational efficiency, and allows customization for different farming types and field conditions.
4. Livestock Health and Behavior Monitoring: Wearable IoT collars and ear tags track livestock vitals such as body temperature and rumination patterns. The integration of Edge AI allows instant anomaly detections like estrus, fever, or reduced activity that can alert farmers in real time. Advantech has implemented an AI-based early detection system for livestock using Edge AI and infrared vision [17]. The system shown in Figure 9.6 deploys a USB 3.0 infrared camera connected to an Edge AI device to continuously monitor the body temperature of each cow. This setup enables real-time detection of elevated temperatures,

Figure 9.5 Autonomous agricultural tractors [16].

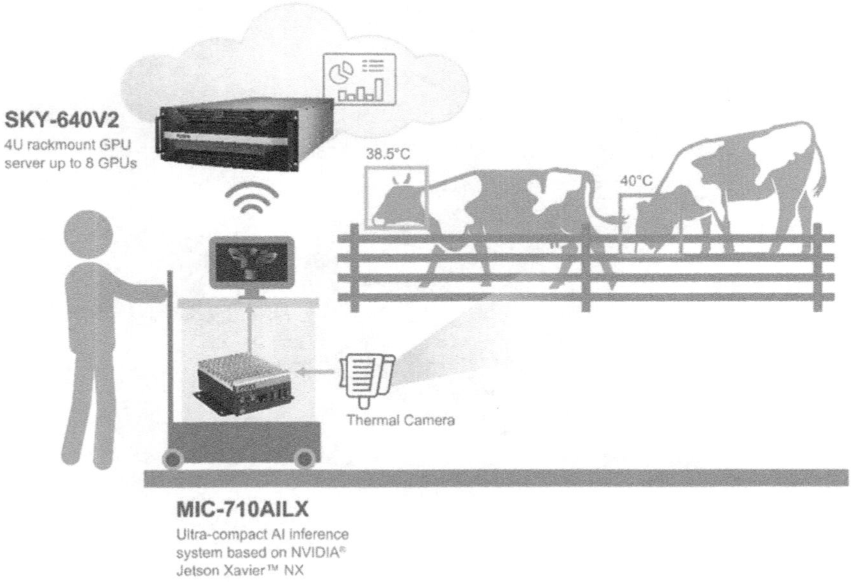

Figure 9.6 The system diagram of AI livestock temperature monitoring [17].

allowing for prompt quarantine and intervention to prevent disease spread. Y.G. Lee et al. proposed the automated feeding systems to dispense customized rations and can boost milk production by 9% in dairy farms [18]. These approaches exemplify how Edge AI and IoT can enable early health issue detection and feed intake optimization, supporting better animal welfare and farm biosecurity.

5. Predictive Analytics for Yield Optimization: GAO Tek has implemented various predictive analytics applications based on Edge AI across several domains [19]. In precision agriculture, for instance, precision planting systems that use edge computing allow farmers to analyze real-time inputs from IoT field sensors, covering factors such as soil moisture, temperature, humidity, and crop condition. Figure 9.7 shows the smart tractors equipped with edge computing capabilities. Processing data directly at the edge allows farmers to act immediately and make better decisions on irrigation, fertilization, and pest control without relying solely on cloud transmission. In the context of supply chains, GAO Tek's edge computing improves efficiency by enabling real-time tracking of perishable products during both transport and storage. IoT-based sensors measuring temperature, humidity, and vibration collect information continuously at different points along the supply chain. IoT devices equipped with temperature, humidity, and vibration sensors continuously gather data at multiple stages along

Figure 9.7 GAO Tek's precision planting system [19].

the supply chain. By analyzing this information locally at the edge, the system can quickly identify any deviations from optimal storage or transport conditions. This immediate detection enables proactive interventions to prevent spoilage and maintain product quality, ensuring that perishable goods arrive safely and in the best possible state.

6. Smart Greenhouses: The Laiwu Tropical Fruit Cultivation Project at Kebai Smart Agriculture Industrial Park in Shandong Province: As illustrated in Figure 9.8, China [20] has implemented SenseCAP LoRaWAN sensors together with edge gateways in greenhouses to track key environmental factors, including CO_2 levels, light intensity, soil temperature and moisture, electrical conductivity, and photosynthetically active radiation (PAR). The edge gateway locally processes sensor data to make immediate control decisions—such as triggering irrigation, fertilization, ventilation, and thermal blanket systems—without relying solely on cloud connectivity. This automation enables precise microclimate management and resource optimization, tailored to the needs of tropical fruit crops being cultivated in a northern climate. The project highlights how IoT and Edge AI technologies can create sustainable, efficient, and location-independent greenhouse operations, significantly reducing labor costs and improving crop outcomes through real-time, autonomous environmental control.

(a) (b)

Figure 9.8 (a) Edge gateway and (b) greenhouses at Kebai Smart Agriculture Industrial Park [20].

9.11 SUMMARY

The integration of IoT and Edge AI is fundamentally reshaping modern agriculture, enabling a transition from traditional, reactive practices to data-driven, proactive, and precision-oriented farming. By leveraging interconnected sensors, autonomous machinery, and real-time analytics, farmers are able to make more efficient use of resources, support better crop and animal health, and increase transparency within the supply chain. These technologies facilitate immediate, localized decision-making, reduce latency, and minimize reliance on continuous internet connectivity—benefits that are particularly valuable in remote or resource-constrained environments. Case studies demonstrate tangible gains in productivity, sustainability, and operational efficiency across diverse agricultural contexts.

Despite these advances, several challenges remain, including hardware limitations, data heterogeneity, connectivity constraints, security and privacy risks, economic barriers, and the need for specialized technical expertise. Addressing these issues will require ongoing innovation in device design, AI model optimization, connectivity infrastructure, and user training. Supportive policy frameworks and accessible financing mechanisms will also be essential to ensure equitable adoption, especially among smallholder and resource-limited farmers.

Looking ahead, the continued evolution of IoT and Edge AI—augmented by developments such as 5G connectivity, blockchain integration, and advanced decision support systems—will further empower farmers to meet the demands of a growing global population while minimizing environmental impacts. The widespread adoption of these technologies is poised to create a more efficient, resilient, and sustainable agricultural system, positioning smart farming as a cornerstone of global food security and environmental stewardship in the 21st century.

REFERENCES

1. A. H. Abdul Hussein, K. A. Jabbar, A. Mohammed, and H. M. Al-Jawahry, "AI and IoT in farming: A sustainable approach," *E3S Web Conf.*, vol. 491, p. 01020, 2024, ICECS'24, doi: 10.1051/e3sconf/202449101020
2. M. El Jarroudi, A. L. Kouadio, P. Delfosse, C. H. Bock, A.-K. Mahlein, X. Fettweis, B. Mercatoris, F. Adams, J. M. Lenné, and S. Hamdioui, "Leveraging edge artificial intelligence for sustainable agriculture," *Nat. Sustain.*, vol. 7, pp. 846–854, 2024, doi: 10.1038/s41893-024-01352-4
3. A. A. Hassan, H. O. Abdullahi, A. F. Ali, and M. H. Ahmed, "Internet of Things in agriculture: A systematic review of applications, benefits, and challenges," *J. Syst. Manag. Sci.*, vol. 14, no. 9, pp. 67–80, 2024, doi: 10.33168/JSMS.2024.0905.
4. E. Bowen, "Emerging trends in the agriculture IoT market 2024," *Telnyx*, Jan. 3, 2024. [Online]. Available: https://telnyx.com/resources/agriculture-iot-market
5. A. A. Chaudhry, R. Mumtaz, S. M. Hassan Zaidi, M. A. Tahir, and S. H. Muzammil School, "Internet of Things (IoT) and Machine Learning (ML) enabled Livestock Monitoring," *2020 IEEE 17th Int. Conf. Smart Communities: Improving Quality of Life Using ICT, IoT and AI (HONET)*, Charlotte, NC, USA, 2020, pp. 151–155, doi: 10.1109/HONET50430.2020.9322666.
6. L. Umutoni and V. Samadi, "Application of machine learning approaches in supporting irrigation decision making: A review," *Agric. Water Manag.*, vol. 294, Art. no. 108710, 2024, doi: 10.1016/j.agwat.2024.108710.
7. A. H. Abdul Hussein, K. A. Jabbar, A. Mohammed, and L. Jasim, "Harvesting the future: AI and IoT in agriculture," *E3S Web Conf.*, vol. 477, Art. no. 00090, 2024, doi: 10.1051/e3sconf/202447700090.
8. I. Ünal and M. Topakci, "Design of a remote-controlled and GPS-guided autonomous robot for precision farming," *Int. J. Adv. Robot. Syst.*, vol. 12, no. 12, pp. 1–10, 2015, doi: 10.5772/62059.
9. A. Rowsell, "Edge computing for agriculture and smart farming," *Simply NUCBlog*. [Online]. Available: https://simplynuc.co.uk/blog/edge-computing-for-agriculture-and-smart-farming/. [Accessed: May 7, 2025].
10. P. Singh, A. Kaur and S. S. Gill., "Machine learning for cloud, fog, edge and serverless computing environments: Comparisons, performance evaluation benchmark and future directions," *Int. J. Grid Util. Comput.*, vol. 13, no. 4, pp. 447–457, 2022.
11. F. Ali, W. Tariq, A. Razzaq, A. Rehman, S. Sarfraz, N. A. Rajput, S. Ali, K. Fatima, S. Jameel, N.Liaqat and Z. Akash, "From sensors to insights: The fusion of AI, edge computing, and precision agriculture," *J. Agric. Biol.*, vol. 3, no. 1, pp. 160–170, 2025, doi: 10.55627/agribiol.003.01.1067.
12. S. M. H. Ifty, B. Hossain, Md R. Ashakin, M. I. Tusher, R. H Shadhin, J. Hoque, R. Chowdhury and A. R. Sunny , "Adoption of IoT in agriculture – Systematic review," *Appl. Agric. Sci.*, vol. 1, no. 1, pp. 1–10, Art. no. 9676, 2024.
13. "How IoT and AI are revolutionizing fungal crop disease management for farmers," *Bivocom*. Available: https://www.bivocom.com/blog/iot-ai-fungal-crop-disease. [Accessed: May 7, 2025].

14. "Aetina, Revolutionizing agriculture with Edge AI," *Aetina Corporation*, May 7, 2025. [Online]. Available: https://www.aetina.com/about-news-detail.php?i=985. [Accessed: May 7, 2025].
15. M. S. Munir, I. S. Bajwa, A. Ashraf, W. Anwar, and R. Rashid, "Intelligent and smart irrigation system using edge computing and IoT," *Complexity*, vol. 2021, Art. no. 6691571, 2021, doi: 10.1155/2021/6691571.
16. "Autonomous agriculture tractors," *Neousys Technology*. [Online]. Available: https://www.neousys-tech.com/en/case-studies/246-autonomous-agriculture-tractors. [Accessed: May 7, 2025].
17. "AI-based early detection system for livestock health management in dairy production," *Advantech*, Apr. 21, 2023. [Online]. Available: https://www.advantech.com/en/resources/case-study/c7b19410-8a40-4231-8d6e-f35bd9cbb811. [Accessed: May 7, 2025].
18. Y.-G. Lee, K. Han, C. Chung, and I. Ji, "Effects of smart farming on the productivity of Korean dairy farms: A case study of robotic milking systems," *Sustainability*, vol. 16, no. 22, Art. no. 9991, 2024, doi: 10.3390/su16229991
19. "Applications of edge computing for IoT in the agriculture, forestry, fishing and hunting industry," *GAO Tek*, May 8, 2025. [Online]. Available: https://gaotek.com/applications-of-edge-computing-for-iot-in-the-agriculture-forestry-fishing-and-hunting-industry/. [Accessed: May 8, 2025].
20. "Smart agriculture: SenseCAP LoRaWAN sensors and gateways enable tropical fruit cultivation in Northern China," *Seeed Studio*, Mar. 10, 2023. [Online]. Available: https://www.seeedstudio.com/blog/2023/03/10/smart-agriculture-sensecap-lorawan-sensors-and-gateways-enable-tropical-fruit-cultivation-in-northern-china/. [Accessed: May 7, 2025].

Chapter 10

Artificial intelligence and Internet of Things in sustainable agriculture

Sumathra Manokaran, Shreya Girish Karkera, M. D. Yaana Muthamma, Srijeev R. Rajesh, and Manjunatha Reddy A. H.

10.1 INTRODUCTION

In the fourth industrial revolution (Industry 4.0), new technologies have emerged to integrate the digital and physical worlds, including cloud computing, automation systems and artificial intelligence (AI). Creating a cross-platform workspace between machines, infrastructures and digital platforms is one of the advantages of Industry 4.0 [1,2]. Crop growing requires a lot of time, money and physical labor. Modern agriculture has several problems like pesticide use, environmental changes and soil, air and water getting worse because of pollution. The use of smart farming technologies is considered a good option to solve these problems and improve the sustainability and productivity of the agribusiness sector [3]. Agricultural methods that improve the use of non-renewable resources, support the expansion and preservation of natural resources and safeguard the environment are referred to as sustainable farming [4]. It aims at producing agricultural products at the lowest possible environmental cost so as to guarantee the food for the next generations [5]. Precision agriculture has become a major innovator that assists farmers in getting maximum crop yields and resource usage via Global Positioning System (GPS), Geographic Information System (GIS) and remote sensing utilization [6]. Decision-making in farming has been changed by the deployment of the Internet of Things (IoT) and AI. The introduction of smart sensor-based real-time crop health, weather and soil moisture monitoring enhances productivity and also minimizes waste [7]. Technology of drones and remote sensing make it possible for big agricultural fields to be viewed from a bird's eye perspective, thus enabling the rapid execution of necessary tasks [8]. The use of drones and satellite images is the trend nowadays for the early identification of pest infestations and nutritional deficiencies. Besides that, blockchain technology is used in order

 DOI: 10.1201/9781003589730-10

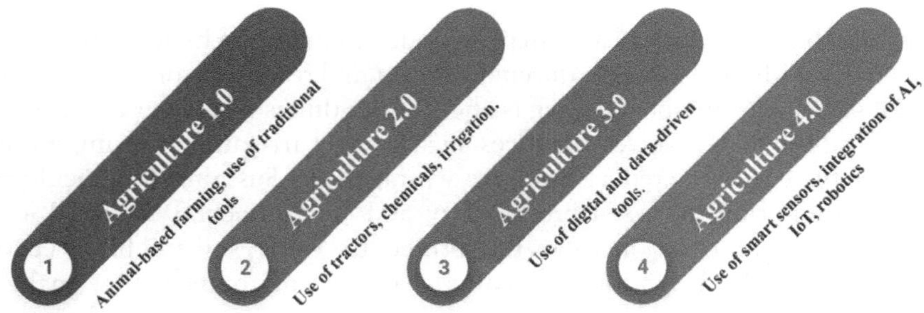

Figure 10.1 Evolution of agriculture from 1.0 to 4.0. From traditional, animal-based farming (1.0) to mechanization and chemical use (2.0), the adoption of digital tools (3.0) and finally smart technologies that incorporate AI, IoT and robots (4.0), the figure illustrates the transformation of agriculture.

to guarantee food safety and fraud reduction [9]. Figure 10.1 shows the evolution of agriculture from 1.0 to 4.0.

Mechanization in agriculture through automated tractors and harvesters substitutes manual labor and brings about a higher efficiency, which is achievable through the use of technology [10]. If farmers have access to real-time data and mobile apps, they can be empowered by weather forecasts, market prices and expert advice [11]. The expected demand for sustainable and health-promoting food production is quite hard to meet. One of the major sustainable agriculture goals is to increase crop yields without contributing to climate change and protecting agro-ecosystems [12,13]. If we want to provide food for the rapidly growing population without harming the environment, sustainable agriculture is a must. Sustainable farming methods ensure long-term food security through promoting resilience and production [14]. AI and machine learning (ML) models are engaged in forecasting insect outbreaks and sourcing the optimal combination of resources like water, fertilizer and pesticides [15]. AI's capacity for rapid and reputable evaluation of huge amounts of data has allowed decision-makers to achieve higher quality decisions in various fields [16]. Besides that, AI also has a positive effect on sustainable growth. It leads to more efficient use of resources and energy, less waste and it facilitates environmental monitoring [17]. One of the examples of food security being ensured is the use of AI-driven intelligent agricultural systems, which assists with water management and forecasting of crop yields [15]. IoT is a new framework that makes it possible for sensors and electronic equipment to communicate with each other online, improving people's quality of life [18]. Traditional living has been replaced by a high-tech lifestyle because of the IoT [19].

10.1.1 Sustainable agriculture using AI and IoT

Sustainable agriculture is a strategy intended to maintain high levels of productivity without damaging the environment and resources for future generations [20]. Sustainable farming is about using things like water and energy efficiently, which involves practices such as drip irrigating and mixing in renewable energy to reduce resource wastage [21]. Sustainable agriculture is an equilibrium between productivity and environmental care and it is also the use of farming methods that protect the water, soil and biodiversity and additionally it is a quite dynamic learning environment that enables innovation, best practice exchange and which also encourages collaboration between farmers, scientists and policymakers for the development of solutions that correspond to a particular context [22]. The addition of AI and the IoT is present as an opportunity to change the whole concept of farming beyond recognition. Through the blend of cognitive data mining (AI) with the sensor networks and device networks (IoT), farmers are enabled to check the situation on the spot, manage the inputs precisely and also carry out many operations automatically. As an example, AI-based systems are capable of forecasting the occurrence of crop disease and deciding on the best harvest time and IoT sensors can be continuously recording nutrient levels, soil moisture and weather [23]. These technologies bring data-centric farming by optimizing resource usage, reducing waste, increasing yield accuracy and being resilient to climate changes [24]. This concept has gained critical importance because traditional farming practices have reached limits. Intensive agriculture has significantly increased yields over the past century, but often at the cost of soil erosion, depleted aquifers and excess nutrient runoff [25]. Figure 10.2 shows AI's role in modern agriculture.

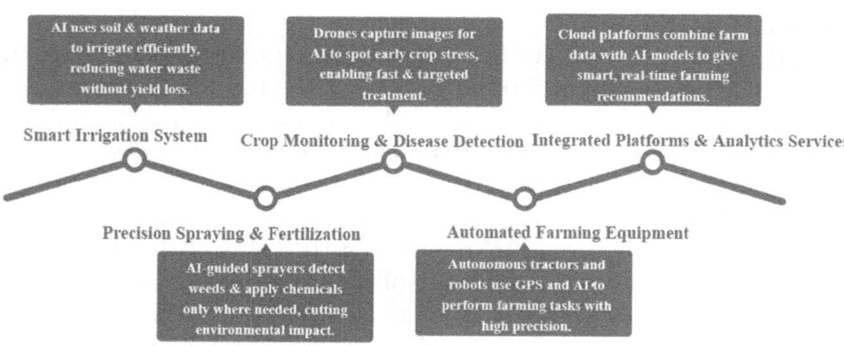

Figure 10.2 AI's role in modern agriculture. AI has several uses, from smart irrigation and crop monitoring to autonomous automation and optimal fertilization. All these technologies improve farming methods' environmental sustainability and production.

10.1.2 Challenges and opportunities in agriculture

Agriculture today faces enormous problems and also holds huge opportunities. Speaking of the difficulties, climate change has an impact whereby floods and droughts become more frequent, hence decreasing yield stability in agriculture. In fact, in most developing regions, infrastructural shortages (poor connectivity, lack of electricity) complicate the use of advanced technology [26]. The interfacing of AI and IoT can address many systemic issues. Interfacing AI and IoT can solve several systemic issues. For example, intelligent farming can make the whole agricultural value chain from planting to harvesting to processing more efficient and sustainable [27]. Farmers who utilize AIoT technologies can enhance production and quality (and therefore gain), as well as lower input costs. This can potentially stabilize markets and enhance food security. Further, the agri-tech sector can grow as start-ups and new companies create sensors, drones, robots and farming analytics platforms. Hence, AI and IoT in farming is not only reshaping the agricultural landscape but is also instrumental in steering it toward a more sustainable, productive and resilient future [28].

10.1.3 Role of digital transformation in agriculture

Digital transformation in the agriculture sector implies utilization of such technologies as big data, cloud computing, IoT and AI to monitor the farm and boost its productivity. These developments emphasize the importance of digital transformation in the development of globally resilient, efficient and sustainable agricultural systems, assuming, however, that there is sufficient infrastructure and that the farmers possess digital skill [29]. Through the growth of innovation ecosystems, the reduction of information asymmetry and the dissemination of knowledge across geographical boundaries, the digital economy increases agricultural productivity. Additionally, through improving market connections, reducing transaction costs and fostering transparency through e-commerce and digital traceability tools, digital transformation has simplified agri-food supply chains [30]. Figure 10.3 shows key elements of technology for smart agriculture.

Sensors and actuators have been better optimized by employing IoT and AI in traditional irrigation-based farming methods, allowing better decision-making and predictive crop water needs. Enhanced crop plot management and water-use efficiency have been demonstrated by irrigation communities using an open and interoperable IoT-based platform [31]. Drones, intelligent sensors and satellite imagery are some examples of smart farming technologies that have appeared due to the rapid digitization of agriculture. These technologies allow for proactive disease and pest surveillance and yield prediction [32].

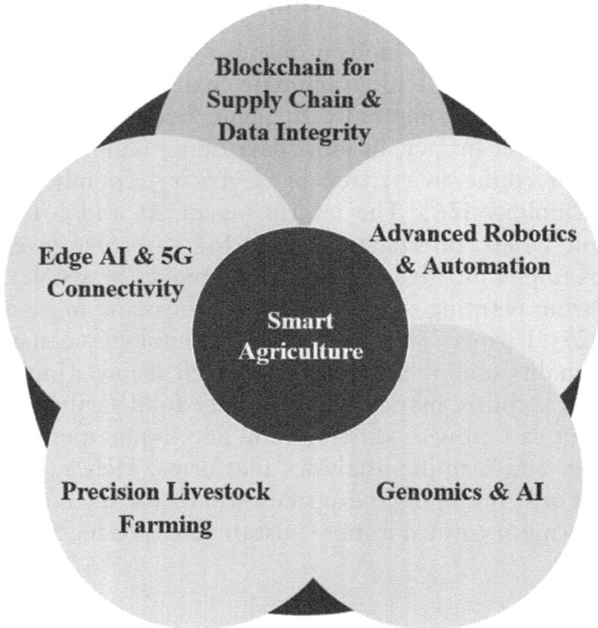

Figure 10.3 Key elements of technology for smart agriculture. The figure depicts the key technologies that power smart agriculture, such as blockchain, robotics, AI, 5G, genomics and precision livestock farming, with an objective to increase the sustainability, productivity and traceability of modern agricultural systems.

10.2 AI AND IoT'S ROLE IN SUSTAINABLE AGRICULTURE

10.2.1 AI's role in sustainable agriculture

AI systems can apply the knowledge of past along with current data (temperature/humidity/soil) to forecast crop yield and nature of weather, and thus, the farmer can be informed in a better way when deciding to sow or harvest [33]. Utilizing the detailed calculation of the quantity of water, fertilizer and pesticides needed in different parts of the field, AI escalates precision agriculture by lessening runoff and decreasing overuse [34]. Computer vision and AI models can also be used by drones or satellites to capture images and then those images are analyzed so that plant diseases and pest infestations can be identified at the earliest stage and consequently the treatment will be faster and the chemicals will be reduced [35]. AI-powered decision support systems (DSS) mix together crop models, soil sensor data and weather predictions to find out the most suitable crop varieties, irrigation schedules and nutrient management techniques [36]. Figure 10.4 shows innovative farming methods in sustainable agriculture using AIoT.

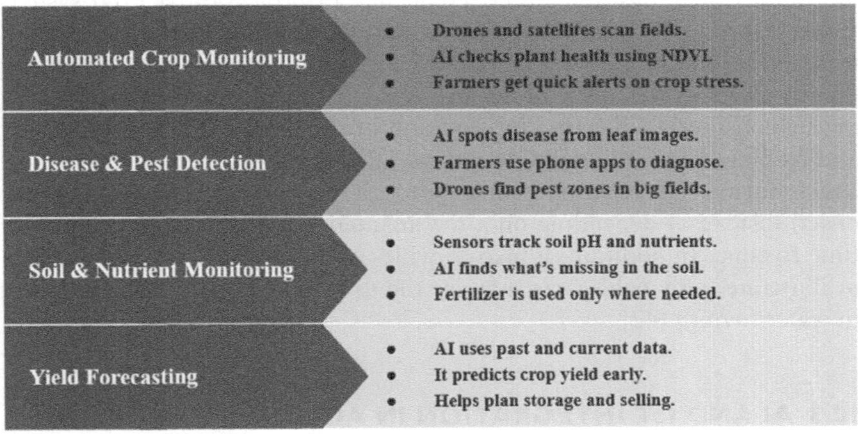

Figure 10.4 Innovative farming methods in sustainable agriculture using AIoT. The figure lists four major uses of technology in agriculture and describes how sensors, drones, satellites and AI evaluate data to enhance farming productivity and decision-making. Each section shows how technology improves farming's decision-making and productivity.

10.2.2 IoT's role in sustainable agriculture

IoT refers to a network of devices, sensors and software that communicate with one another. It is a system that gathers and exchanges data from sensors found everywhere at once. IoT makes sustainable agriculture possible that relies on data to make decisions that enhance the yield, save waste and adopt eco-friendly farming methods. The use of IoT in sustainable agriculture enables the farmers to be able to monitor their farms in real time, utilize the resources to the optimal extent and make choices based on the information that they get. All these are the central aspects of sustainable agriculture that can be obtained with the minimum harm to the environment [37,38]. IoT sensors offer farmers instantaneous measurements of pH, nutrient levels, soil moisture and temperature to use as non-polluting and optimized use of the inputs for fertilization and irrigation. IoT-based irrigation relies on readings from soil and weather stations to adjust the water consumption schedule automatically in accordance with the prevailing conditions, thereby conserving water and energy [39]. Smart IoT devices attached to animals monitor their health, store all the movements and temperature and will inform exactly where they are at all times, so animal welfare will increase, and if a disease occurs, it will be identified at the initial stage [40]. Without the requirement for human interaction, IoT, from sensor feedback, enables optimization of greenhouses growth conditions, by automatically controlling lighting, humidity, ventilation and

CO_2 levels. IoT tags and Radio Frequency Identification (RFID) systems transmit produce's location, humidity and temperature during transportation and storage, which is the main reason for reducing waste and ensuring quality of food [41]. To sum up, IoT is the nervous system of smart farming by constantly collecting information of the field, reliably sending data and feeding data into analysis engines. Through real-time monitoring, IoT assists farmers in reacting promptly to alterations (such as frost or pest onset) instead of depending on a few manual inspections that occur from time to time. In addition, it also provides accurate resource management: to illustrate, only field parts that are the driest may be irrigated, resulting in water savings [42].

10.3 AI AND IoT INTEGRATION IN AGRICULTURE

Agriculture is undergoing a revolution due to the combination of AI and the IoT, or AIoT, which makes it possible for intelligent, automated and real-time decision-making systems. While IoT offers continuous data collection from physical environments like farms, fields and greenhouses, AI offers advanced analytics, pattern recognition and decision support [43]. AIoT supports the core principles of sustainable agriculture by reducing environmental damage, optimizing resource use and enhancing food security. IoT sensors, including soil moisture sensors, temperature sensors and weather stations, collect large amounts of data in real time. Farmers can make informed decisions about irrigation, fertilization and harvest timing by utilizing AI algorithms to analyze this data and generate actionable insights [7]. The goal of precision agriculture is to deliver the proper care at the proper time and location. AI analyzes micro-level data gathered by IoT like localized rainfall or nutrient levels to provide personalized solutions. For example, AI algorithms guide self-driving machines to spray pesticides or fertilizers where they are needed only, reducing environmental pollution, utilizing GPS-labeled field maps and sensor feeds [36]. AIoT systems track agricultural equipment by utilizing embedded IoT sensors that record temperature, vibration and usage time. AI programs then process this information to forecast equipment breakdown and facilitate early repair. This encourages more efficient and sustainable farm practices by reducing equipment downtime, conserving funds and prolonging asset life [44]. Figure 10.5 shows AIoT's effects on various aspects of farming.

AI models trained on image datasets can detect early signs of plant disease, nutrient deficiencies or pest infestations which reduces the need for large-scale pesticide use and helps improve yield forecasts, leading to both ecological and economic sustainability [33]. AIoT also improves post-harvest handling and logistics. IoT sensors monitor storage temperature, humidity and location in transit. AI software applies this information to

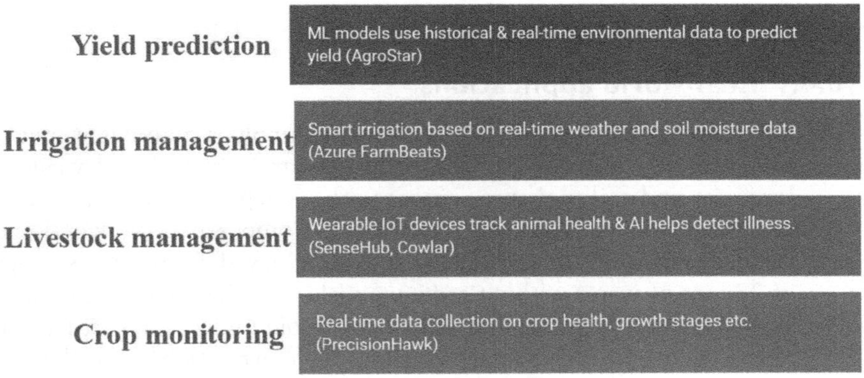

Figure 10.5 AIoT's effects on various aspects of farming. Four major areas of agricultural technology are depicted in the figure: crop monitoring, livestock management, irrigation management and yield prediction.

optimize routes and storage procedures, reducing spoilage and maintaining food quality in transit [42]. Several AIoT systems are incorporated into farm management systems that are user-friendly. Real-time alerts for major events, like infestations by insects or frost damage, are provided for farmers [45]. To forecast yields and identify limiting factors, AI systems analyze sensor and satellite inputs. Comparing weather patterns to past infestations, they can forecast pest or disease outbreaks ahead of time. Farmers can subsequently make timely countermeasures [46]. Sustainability is increased since smart technology takes fewer inputs. Through exact use of only the amount of water or pesticide required, AI-IoT agriculture saves resources and minimizes environmental effect [47]. Since robots can weed or harvest automatically through AI vision and drones can plant seeds, scout out regions or even spray crops without pilots, farmers can view their farms remotely from their smartphones. Thus, farmers can now take care of broader areas or devote themselves to planning strategically because of AI and IoT [48]. IoT and AI transform intuitive choices into evidence-based plans. Predictive modeling is enabled by the integration of continuous sensor feeds and past farm data. Yield prediction models, for instance, use plant wellness and weather data to inform market planning and supply chain decisions [49]. Predictive analytics can alert growers to potential issues (such as nutrient deficiencies) before signs appear. Agronomists and policymakers over time may derive benefit from regional trends (climate trends, soil changes) as exposed when various farms' data are aggregated. Farming in general is made clearer and easier to control when AI analytics and IoT data collaborate [50].

10.4 REAL-WORLD APPLICATIONS AND CASE STUDIES

10.4.1 Real-world applications

a. Smart Irrigation Systems: A prominent example is automated irrigation using AI and IoT. In one study, researchers deployed weather forecasts and soil moisture sensors with a deep reinforcement learning (RL) model to irrigate a rice field. Compared to conventional flooded irrigation, the AI-driven system achieved an average savings of 23 mm of water without reducing yield. The RL model learned to exploit rainfall forecasts and soil data to avoid unnecessary irrigation. Such systems are increasingly adopted: smart irrigation controllers on many farms now schedule watering events based on real-time sensor data rather than timers, greatly reducing water use [51].
b. Automated Farming Equipment: Autonomous tractors and robots are becoming more prevalent. These IoT-connected devices use AI for navigation and task execution. A driverless tractor, for instance, might be equipped with sensors to monitor the soil as it moves and use ML and GPS to sow or plow with centimeter accuracy. Robotic arms are used to pick strawberries, and AI vision is used to identify ripe fruit. Despite their infancy, these applications are used for labor-intensive crops. Their development is guided by the same AI-IoT principles of data-driven control highlighted in research [52].
c. Integrated Platforms and Analytics Services: Many agricultural tech companies now offer cloud platforms that integrate farm data and provide AI analytics. Farmers can upload sensor data (soil, weather, equipment) to a centralized system that runs ML models to produce actionable advice. These platforms frequently provide dashboards with field condition maps and anomaly warnings. Some systems use AI algorithms, for instance, to recommend planting densities or crop rotations every season based on a farmer's past production and soil data. While the details vary, the key is that AIoT solutions are transitioning from prototypes to practical tools on working farms [53].

10.4.2 Case studies

a. Wangree Plant Factory – Smart Vertical Farming (Thailand): Wangree Health Factory Company in Thailand utilizes a vertical farming system combined with IoT technologies to produce organic vegetables and fruits. The system automates watering, lighting, nutrient addition and temperature control, allowing year-round high-quality production under controlled environments [1].
b. CropIn – Smart Farm Management (India): CropIn Technology Solutions provides real-time weather, soil and crop data using AI and

Table 10.1 AI/IoT platforms and tools for agriculture

Sr. no.	*Platform/tool*	*Technologies used*	*Applications*
1.	John Deere Operations Center	IoT Sensors, AI, GPS	Crop lifecycle monitoring, satellite-guided operations and smart tractors [15]
2.	IBM Watson Decision Platform	Cloud Computing, AI, ML	Weather forecasting, yield optimization, and crop disease prediction [56]
3.	SenseFly eBee Drones	Drone Tech, Multispectral Imaging	Mapping farm health using high-resolution aerial imagery [57]
4.	PrecisionHawk	Computer Vision, Deep Learning, UAV (Drone)	Drone-based crop imaging and analytics, plant health monitoring [58]
5.	SmartFarm (India)	AI, Satellite Imaging, Mobile Apps	Farmer advisory, soil nutrient analysis, and pest alerts [59]

Five smart agriculture platforms and tools are included in the table, along with the technology and uses for each.

IoT-based farm management tools. Millions of acres in India have seen the implementation of their ideas, which have assisted farmers in lowering input prices and increasing output [54]. Table 10.1 shows AI/IoT Platforms and Tools for Agriculture.

c. Orchard Modernization: Project PANTHEON is focused on the cultivation of hazelnuts using a SCADA (Supervisory Control and Data Acquisition) system. For automating fertilizing and watering schedules, it employs real-time sensor data for temperature, moisture levels and other variables. This encourages healthier, more productive trees as well as improving water and nutrient usage [55].

10.5 AI AND IoT-BASED SUSTAINABLE FARMING PRACTICES

A major leap forward in sustainable agriculture is the integration of AIoT technologies. In this chapter, we discuss three important areas: precision agriculture and automation, smart irrigation and crop monitoring and disease management.

10.5.1 Automation of agriculture and precision farming

Precision agriculture refers to farm management approaches that optimize field-level inputs (seeds, water, nutrients) and operations (planting, spraying) based on the variability within a field. Use of AIoT in precision farming enhances efficiency and productivity.

a. Variable-Rate Technology (VRT): IoT sensors and GPS mapping allow fields to be divided into zones with different characteristics (soil type, moisture). AI algorithms process this data to create variable-rate maps. Modern tractors equipped with IoT can adjust seed drop or fertilizer flow on-the-fly, following these maps. This maximizes productivity in good areas and conserves inputs in poor areas. Models of AI can learn from past yields to refine these zones each season. The result is increased yield consistency and lower input costs [60].
b. Autonomous Machinery: Automated farming equipment is a key innovation. AI-powered GPS-guided tractors need little human supervision. Robotic weeding systems remove weeds with precision tools by using cameras and convolutional neural networks to identify crops from weeds. Produce can be selected by automated harvesters using ripeness estimates. Compared to people, these systems are more efficient and accurate at labor-intensive jobs. AI, for instance, can be used by an autonomous vineyard robot to identify the best times to select grapes and harvest them. All these machines rely on IoT connectivity to receive updates and coordinate with farm management systems [52].
c. Data-Driven Crop Management: Farmers can now use AI models to choose crop kinds, arrange crop rotations and maximize planting times. Through the analysis of multi-year IoT data on soil nutrients, weather and yields, ML models have the ability to forecast which crop will thrive in each field under the conditions of the upcoming season. Decision support apps use AI to integrate agronomic rules with real-time data, providing recommendations for operations (e.g., plant corn this week; soil nutrients are optimal) [61].

10.5.2 Automatic water management and irrigation systems

Smart irrigation systems use water wisely by combining weather and soil sensors with AI control.

a. Determining the Moisture Content in Soil: Water levels are consistently monitored by IoT soil moisture sensors in various zones of the field. AI processing applies those data and weather predictions to decide on the precise volume and timing of irrigation. Delaying irrigation, for instance, in order to prevent overwatering, the system can delay watering if rain is forecasted. Targeted irrigation may be initiated, however, if sections of the field become dry. This significantly reduces wasteful watering [62].

b. Automated Control of the Valve: I-controlled actuated valves may be opened or closed to supply water zone by zone. In a standard arrangement, central water sources (such as canals or pumps) supply several intelligent valves. The AI schedules the valves to function, using water just where sensors show shortage. Some systems even use satellite evapotranspiration data to improve irrigation scheduling [63].
c. RL Irrigation: Sophisticated research uses RL, an AI system that learns through trial and error, to optimize irrigation. For example, one tested a deep-Q-learning model using decades' worth of weather and growth history for rice crops. In simulation, this RL system learned irrigation strategies that conserved an average of 23 mm of water over conventional methods without sacrificing yield. Through ongoing learning from results, RL controllers become more skilled at reconciling crop requirements with conservation objectives [64,65].

In total, smart irrigation illustrates how IoT gives the sensor data (weather, soil moisture) and how AI provides the decision engine. The end result is a more sustainable water management. Water is supplied in the right quantity, at the right time, to the right place due to AI control and real-time sensor monitoring, making irrigation adaptive. In dry areas particularly, such systems are revolutionizing water use in agriculture.

10.5.3 Crop monitoring, detecting diseases and forecasting yield

Sustainable yields depend on crop health maintenance and performance forecasting. AI and IoT significantly improve these functions.

a. Automated Crop Monitoring: Today, IoT cameras and sensors are mounted on aerial drones and satellites that regularly patrol fields. These streams of image data, when processed by AI, can track plant health indicators (such as NDVI, normalized difference vegetation index) over extensive areas. For instance, a decline in NDVI in a part of a field can signal incipient stress. Automating the monitoring process, farmers get alerts much quicker than with field scouting alone [66].
b. Detection of Diseases and Pests: Visual signs of disease or insect damage on leaves is detectable with ML models that have learned from image datasets. A Deep Learning model analyzes the images captured by IoT devices, like fixed cameras or phone attachments. Camera-equipped drones scan for abnormalities in fields and point out areas needing attention for large farms. This automated identification allows for targeted intervention before an infestation is widespread [67].

c. Yield Prediction: By incorporating weather forecasts, past yields and current IoT sensor readings into AI models, one can make precise predictions about upcoming yield. Precise yield projections enable improved supply chain planning (storage, marketing) as well as policy formulation (food security initiatives). They also enable detection of which fields are not performing well, leading to corrective actions in future seasons [68].

In all of these fields, the secret is ongoing data streaming from the farm (IoT) and strong pattern detection (AI).

10.6 LIMITATIONS AND CHALLENGES OF AI AND IoT IN SUSTAINABLE AGRICULTURE

10.6.1 Data privacy, security and ethics

Field maps, sensor logs and usage trends are among the vast amounts of data produced by smart farms. This data can be commercially sensitive. Cloud-based data storage makes it susceptible to illegal use or cyber-attacks. Safeguarding farm data requires robust encryption and clear agreements on data ownership. There are some ethical questions, for example, if AI-driven decisions fail, who is liable, the farmer or the tech provider. Without careful handling of privacy and ethics, farmer adoption of AI-IoT may be slow [69,70].

10.6.2 Cost, infrastructure and accessibility barriers

The absence of electricity and internet connectivity in many places makes it challenging for many farmers to start using technologies like AI, IoT sensors, etc. Example: if a soil sensor is unable to deliver its data to the cloud, it is ineffective. Also, in less developed areas, farmers often will not have access to computers and smartphones. All these can hinder equitable access to smart farming benefits [71].

10.6.3 Technical and implementation challenges

AI models often need large amounts of high-quality data to train properly, and in agriculture, collecting labeled data (e.g., images of diseased plants with exact diagnoses) can be labor-intensive. Sensor drift and calibration are problems since faulty sensors may cause AI to make poor conclusions. In addition to this, farmers and agronomists must be taught how to read AI analytics and service IoT networks. Without education and guidance, even a well-conceived system will not work in practice [69,72].

10.7 FUTURE TRENDS AND INNOVATIONS

10.7.1 Upcoming technologies in AI and IoT for agriculture

a. 5G Networking and Edge AI: Now it is not necessary to push all information to the cloud when AI programs are run locally on field devices (edge computing). An AI-equipped IoT camera, for example, might detect pests on board and simply raise an alarm, saving bandwidth. The advent of 5G networks, which are high-capacity and low-latency, will enable farm devices to coordinate in real time. On large farms, 5G connection in conjunction with IoT will help smooth communication between a vast number of robots and sensors [73].
b. Advanced Robotics and Automation: Under the guidance of AI vision and IoT mapping, future robots may do activities like individual plant care or selective harvesting. Fields could be tended by swarms of tiny robots or drones that communicate with each other over mesh networks. These systems will be dependent on advancements in AI planning algorithms, computer vision, and battery life [74].
c. Blockchain for Data Integrity and Supply Chain: By combining blockchain technology with IoT, it is possible to produce unchangeable records of the origins and methods of food production, improving food safety and tracking of fair trade. In order to provide end-to-end traceability from farm to table, sensors might automatically record cultivation conditions to a blockchain. Given that blockchain can grant farmers control of their data records, this also resolves some ethical issues [75].

10.7.2 IoT and AI's contribution to global food security

With growth in population and climate change, there are concerns about global food security. In order to solve this, AI and IoT will be essential. First, AI-IoT technologies aid in bridging the gap between supply and demand for food by boosting yields and decreasing losses. Better matching of production to demand is made possible by models that predict yields. Harvest failures are avoided by using technology that increases farm resilience (such as pest detection and drought-resistant irrigation). Innovations in digital farming present a viable strategy for feeding the growing global population in a sustainable manner, as one evaluation highlights [76]. The increasing issues of water scarcity and soil erosion are changing due to advances in agriculture. It reduces unnecessary usage of water and preserves essential resources, pointing to a more thoughtful and effective approach to resource management in agriculture. The agricultural industry is also experiencing a revolution in disaster preparedness and policy formulation thanks to AI and IoT technologies. This method facilitates the creation of programs aimed at enhancing long-term food security and strengthens the government's capacity to manage emerging problems. For instance, predictive algorithms might encourage the growth of crops that mature quickly or offer import subsidies if a famine is approaching [77].

10.7.3 Sustainability and agri-tech's future

In addition to production gains, future agri-tech will be evaluated on its effects on the environment and social justice. Sustainability itself will therefore be a fundamental design requirement. Carbon accounting may be incorporated into systems to reduce emissions, and ML may optimize processes for both profit and low ecological footprints [78]. Integrated digital farms are becoming a reality: highly automated locations with satellites, drones, weather stations and soil sensors that all input data to an AI decision support center. Through dashboards that offer specific suggestions for optimizing results within sustainability limits, farmers will engage with these systems. Crop planning, input utilization, harvesting and logistics are all maximized holistically in this concept [79]. Crucially, social and economic aspects will also be included in sustainability goals, such as making sure that farmers equitably benefit from AI-IoT and that it promotes rural development [75].

10.8 SUMMARY

We looked at how AIoT can help with sustainable agriculture in this chapter. These technologies make farms more productive, efficient and resource-conserving by providing cognitive analysis (via AI) and precise monitoring (through IoT sensors). Intelligent farming practices can maximize the agricultural value chain, resulting in more output, less waste and increased efficiency. Precision methods, the growing use of autonomous machinery and improved fertilization are all examples of how AIoT is radically changing agricultural practices. These developments lower resource usage while simultaneously increasing yields. It is impossible to ignore the variety of difficulties brought about by the adoption and integration of such technology. The agri-tech sector appears set to undergo a significant transition with the support of technological advances like Edge AI, robust 5G connectivity, blockchain systems, advanced robotics and genomic analytics. In light of continuous climate change, these advances are expected to be crucial in improving agricultural production and cutting waste, two factors that are critical to addressing global food security [80]. Future research should focus on developing more dependable and affordable IoT devices, making sure data security protocols are adhered to and developing AI models that can operate with less data. In order to find solutions for various real-world farming systems, cooperation between technologists, agronomists and farmers will be essential. In conclusion, AI and IoT are transforming agriculture into a networked, intelligent system that can produce a lot of items with little influence on the environment. In order to steer agriculture toward a robust and sustainable future, these technologies will become very important [81]. By using this technology properly, the world community may work to end hunger and save the environment at the same time.

REFERENCES

1. Santiteerakul, S., A. Sopadang, K. Yaibuathet Tippayawong, and K. Tamvimol. 2020. "The Role of Smart Technology in Sustainable Agriculture: A Case Study of Wangree Plant Factory." *Sustainability* 12 (11): 4640. https://doi.org/10.3390/su12114640.
2. Saunila, M., M. Nasiri, J. Ukko, and T. Rantala. 2019. "Smart Technologies and Corporate Sustainability: The Mediation Effect of Corporate Sustainability Strategy." *Computers in Industry* 108: 178–85. https://doi.org/10.1016/j.compind.2019.03.003.
3. AlZubi, A. A., and K. Galyna. 2023. "Artificial Intelligence and Internet of Things for Sustainable Farming and Smart Agriculture." *IEEE Access* 11: 78686–92. https://doi.org/10.1109/ACCESS.2023.3298215.
4. Chen, Y., Z. Sun, Y. Zhou, W. Yang, and Y. Ma. 2024. "The Future of Sustainable Farming: An Evolutionary Game Framework for the Promotion of Agricultural Green Production Technologies." *Journal of Cleaner Production* 460: 142606. https://doi.org/10.1016/j.jclepro.2024.142606.
5. Sharma, P., P. Sharma, and N. Thakur. 2024. "Sustainable Farming Practices and Soil Health: A Pathway to Achieving SDGs and Future Prospects." *Discover Sustainability* 5 (1): 250. https://doi.org/10.1007/s43621-024-00447-4.
6. Zhang, C., and J. M. Kovacs. 2012. "The Application of Small Unmanned Aerial Systems for Precision Agriculture: A Review." *Precision Agriculture* 13 (6): 693–712. https://doi.org/10.1007/s11119-012-9274-5.
7. Wolfert, S., L. Ge, C. Verdouw, and M.-J. Bogaardt. 2017. "Big Data in Smart Farming – A Review." *Agricultural Systems* 153: 69–80. https://doi.org/10.1016/j.agsy.2017.01.023.
8. Mogili, U. R., and B. B. V. L. Deepak. 2018. "Review on Application of Drone Systems in Precision Agriculture." *Procedia Computer Science* 133: 502–09. https://doi.org/10.1016/j.procs.2018.07.063.
9. Pampuri, A., A. Tugnolo, V. Giovenzana, A. Casson, R. Guidetti, and R. Beghi. 2021. "Design of Cost-Effective LED Based Prototypes for the Evaluation of Grape (*Vitis vinifera* L.) Ripeness." *Computers and Electronics in Agriculture* 189: 106381. https://doi.org/10.1016/j.compag.2021.106381.
10. Bechar, A., and C. Vigneault. 2016. "Agricultural Robots for Field Operations: Concepts and Components." *Biosystems Engineering* 149: 94–111. https://doi.org/10.1016/j.biosystemseng.2016.06.014.
11. Mittal, S., and M. Mehar. 2016. "Socio-Economic Factors Affecting Adoption of Modern Information and Communication Technology by Farmers in India: Analysis Using Multivariate Probit Model." *The Journal of Agricultural Education and Extension* 22 (2): 199–212. https://doi.org/10.1080/1389224X.2014.997255.

12. Khangura, R., D. Ferris, C. Wagg, and J. Bowyer. 2023. "Regenerative Agriculture – A Literature Review on the Practices and Mechanisms Used to Improve Soil Health." *Sustainability* 15 (3): 2338. https://doi.org/10.3390/su15032338.
13. Tahat, M. M., K. M. Alananbeh, Y. A. Othman, and D. I. Leskovar. 2020. "Soil Health and Sustainable Agriculture." *Sustainability* 12 (12): 4859. https://doi.org/10.3390/su12124859.
14. Pretty, J. 2008. "Agricultural Sustainability: Concepts, Principles and Evidence." *Philosophical Transactions of the Royal Society B: Biological Sciences* 363 (1491): 447–65. https://doi.org/10.1098/rstb.2007.2163.
15. Liakos, K., P. Busato, D. Moshou, S. Pearson, and D. Bochtis. 2018. "Machine Learning in Agriculture: A Review." *Sensors* 18 (8): 2674. https://doi.org/10.3390/s18082674.
16. Jordan, M. I., and T. M. Mitchell. 2015. "Machine Learning: Trends, Perspectives, and Prospects." *Science* 349 (6245): 255–60. https://doi.org/10.1126/science.aaa8415.
17. Rolnick, D., P. L. Donti, L. H. Kaack, K. Kochanski, A. Lacoste, K. Sankaran, A. S. Ross, N. Milojevic-Dupont, N. Jaques, A. Waldman-Brown, A. S. Luccioni, T. Maharaj, E. D. Sherwin, S. K. Mukkavilli, K. P. Kording, C. P. Gomes, A. Y. Ng, D. Hassabis, J. C. Platt, … Y. Bengio. 2023. "Tackling Climate Change with Machine Learning." *ACM Computing Surveys* 55 (2): 1–96. https://doi.org/10.1145/3485128.
18. Sfar, A. R., Z. Chtourou, and Y. Challal. 2017. "A Systemic and Cognitive Vision for IoT Security: A Case Study of Military Live Simulation and Security Challenges." In *2017 International Conference on Smart, Monitored and Controlled Cities (SM2C)*, Sfax, Tunisia, 101–5. https://doi.org/10.1109/SM2C.2017.8071828.
19. Kumar, S., P. Tiwari, and M. Zymbler. 2019. "Internet of Things Is a Revolutionary Approach for Future Technology Enhancement: A Review." *Journal of Big Data* 6 (1): 111. https://doi.org/10.1186/s40537-019-0268-2.
20. Hobbs, P. R., K. Sayre, and R. Gupta. 2008. "The Role of Conservation Agriculture in Sustainable Agriculture." *Philosophical Transactions of the Royal Society B: Biological Sciences* 363 (1491): 543–55. https://doi.org/10.1098/rstb.2007.2169.
21. Tilman, D., C. Balzer, J. Hill, and B. L. Befort. 2011. "Global Food Demand and the Sustainable Intensification of Agriculture." *Proceedings of the National Academy of Sciences* 108 (50): 20260–64. https://doi.org/10.1073/pnas.1116437108.
22. Pretty, J. N., A. D. Noble, D. Bossio, J. Dixon, R. E. Hine, F. W. T. Penning de Vries, and J. I. L. Morison. 2006. "Resource-Conserving Agriculture Increases Yields in Developing Countries." *Environmental Science & Technology* 40 (4): 1114–19. https://doi.org/10.1021/es051670d.
23. Bala Murugan, M., M. K. Rajagopal, and D. Roy. 2021. "IoT Based Smart Agriculture and Plant Disease Prediction." *Journal of Physics: Conference Series* 2115 (1): 012017. https://doi.org/10.1088/1742-6596/2115/1/012017.
24. Ashraf, H., and M. T. Akanbi. n.d. "Sustainable Agriculture in the Digital Age: Crop Management and Yield Forecasting with IoT, Cloud, and AI." *Tensorgate Journal of Sustainable Technology and Infrastructure for Developing Countries* 2023 (6): 64–71. https://orcid.org/0000-0002-2419-4626.

25. Abdul Hussein, A. H., K. A. Jabbar, A. Mohammed, and H. M. Al-Jawahry. 2024. "AI and IoT in Farming: A Sustainable Approach." *E3S Web of Conferences* 491: 01020. https://doi.org/10.1051/e3sconf/202449101020.
26. Hundal, G. S., C. M. Laux, D. Buckmaster, M. J. Sutton, and M. Langemeier. 2023. "Exploring Barriers to the Adoption of Internet of Things-Based Precision Agriculture Practices." *Agriculture* 13 (1): 163. https://doi.org/10.3390/agriculture13010163.
27. Kiani, F., G. Randazzo, I. Yelmen, A. Seyyedabbasi, S. Nematzadeh, F. A. Anka, F. Erenel, M. Zontul, S. Lanza, and A. Muzirafuti. 2022. "A Smart and Mechanized Agricultural Application: From Cultivation to Harvest." *Applied Sciences* 12 (12): 6021. https://doi.org/10.3390/app12126021.
28. Alreshidi, E. 2019. "Smart Sustainable Agriculture (SSA) Solution Underpinned by Internet of Things (IoT) and Artificial Intelligence (AI)." *International Journal of Advanced Computer Science and Applications* 10 (5): 93–102 . https://doi.org/10.14569/IJACSA.2019.0100513.
29. Kovács, T., S. Botos, and J. Felföldi. 2023 "What Does Digital Transformation Do in Agriculture? A Systematic Literature Review of Agri-Food Sector's Digitalisation." *Journal of Agricultural Informatics* 14 (2): 26–42. https://doi.org/10.17700/jai.2023.14.2.707.
30. Fu, L., J. Min, C. Luo, X. Mao, and Z. Liu. 2024. "The Impact of Digitalization on Agricultural Green Development: Evidence from China's Provinces." *Sustainability* 16 (21): 9180. https://doi.org/10.3390/su16219180.
31. López-Morales, J. A., J. A. Martínez, and A. F. Skarmeta. 2020. "Digital Transformation of Agriculture through the Use of an Interoperable Platform." *Sensors* 20 (4): 1153. https://doi.org/10.3390/s20041153.
32. Hassoun, A., H. J. P. Marvin, Y. Bouzembrak, F. J. Barba, J. M. Castagnini, N. Pallarés, R. Rabail, R. M. Aadil, S. P. Bangar, R. Bhat, J. Cropotova, S. Maqsood, and J. M. Regenstein. 2023. "Digital Transformation in the Agri-Food Industry: Recent Applications and the Role of the COVID-19 Pandemic." *Frontiers in Sustainable Food Systems* 7: 1217813. https://doi.org/10.3389/fsufs.2023.1217813.
33. Kamilaris, A., and F. X. Prenafeta–Boldú. 2018b. "Deep Learning in Agriculture: A Survey." *Computers and Electronics in Agriculture* 147: 70–90. https://doi.org/10.1016/j.compag.2018.02.016.
34. Zhang, N., M. Wang, and N. Wang. 2002. "Precision Agriculture — A Worldwide Overview." *Computers and Electronics in Agriculture* 36 (2–3): 113–32. https://doi.org/10.1016/S0168-1699(02)00096-0.
35. Mohanty, S. P., D. P. Hughes, and M. Salathé. 2016. "Using Deep Learning for Image-Based Plant Disease Detection." *Frontiers in Plant Science* 7: 1419. https://doi.org/10.3389/fpls.2016.01419.
36. Shamshiri, R., C. Weltzien, I. A. Hameed, I. J. Yule, T. E. Grift, S. K. Balasundram, L. Pitonakova, D. Ahmad, and G. Chowdhary. 2018. "Research and Development in Agricultural Robotics: A Perspective of Digital Farming." *International Journal of Agricultural and Biological Engineering* 11 (4): 1–11. https://doi.org/10.25165/j.ijabe.20181104.4278.
37. Kumari, D. A., A. Obulesh, K. R. Babu, Y. P. Kumar, and K. S. R. K. Sarma. 2024. "Leveraging Internet of Things (IoT) for Sustainable Agriculture: A Comprehensive Review and Future Perspectives." In *Proceedings of the International Conference on Sustainable Technologies*, Amalapuram, India, 587–97. https://doi.org/10.2991/978-94-6463-471-6_57.

38. Wolfert, S., and G. Isakhanyan. 2022. "Sustainable Agriculture by the Internet of Things – A Practitioner's Approach to Monitor Sustainability Progress." *Computers and Electronics in Agriculture* 200: 107226. https://doi.org/10.1016/j.compag.2022.107226. 36
39. Rawal, S. 2017. "IoT Based Smart Irrigation System." *International Journal of Computer Applications* 159 (8): 7–11. https://doi.org/10.5120/ijca2017913001.
40. Talavera, J. M., L. E. Tobón, J. A. Gómez, M. A. Culman, J. M. Aranda, D. T. Parra, L. A. Quiroz, A. Hoyos, and L. E. Garreta. 2017. "Review of IoT Applications in Agro-Industrial and Environmental Fields." *Computers and Electronics in Agriculture* 142: 283–97. https://doi.org/10.1016/j.compag.2017.09.015.
41. Verdouw, C. N., J. Wolfert, A. J. M. Beulens, and A. Rialland. 2016. "Virtualization of Food Supply Chains with the Internet of Things." *Journal of Food Engineering* 176: 128–36. https://doi.org/10.1016/j.jfoodeng.2015.11.009.
42. Bwambale, E., F. K. Abagale, and G. K. Anornu. 2022. "Smart Irrigation Monitoring and Control Strategies for Improving Water Use Efficiency in Precision Agriculture: A Review." *Agricultural Water Management* 260: 107324. https://doi.org/10.1016/j.agwat.2021.107324.
43. Shoba, H., G. Nagaraja, P. N. Krishnamma, and M. S. Sreedevi. 2024. "Smart Farming: Integration of IoT and AI in Agricultural Engineering." *International Journal of Agriculture Extension and Social Development* 7 (9): 692–700. https://doi.org/10.33545/26180723.2024.v7.i9j.1120.
44. Tzounis, A., N. Katsoulas, T. Bartzanas, and C. Kittas. 2017. "Internet of Things in Agriculture, Recent Advances and Future Challenges." *Biosystems Engineering* 164: 31–48. https://doi.org/10.1016/j.biosystemseng.2017.09.007.
45. Muhammed, D., E. Ahvar, S. Ahvar, M. Trocan, M.-J. Montpetit, and R. Ehsani. 2024. "Artificial Intelligence of Things (AIoT) for Smart Agriculture: A Review of Architectures, Technologies and Solutions." *Journal of Network and Computer Applications* 228: 103905. https://doi.org/10.1016/j.jnca.2024.103905.
46. Delfani, P., V. Thuraga, B. Banerjee, and A. Chawade. 2024. "Integrative Approaches in Modern Agriculture: IoT, ML and AI for Disease Forecasting amidst Climate Change." *Precision Agriculture* 25 (5): 2589–2613. https://doi.org/10.1007/s11119-024-10164-7.
47. Zia, H., A. Rehman, N. R. Harris, S. Fatima, and M. Khurram. 2021. "An Experimental Comparison of IoT-Based and Traditional Irrigation Scheduling on a Flood-Irrigated Subtropical Lemon Farm." Sensors 21 (12): 4175. https://doi.org/10.3390/s21124175.
48. Anwarul, S., T. Misra, and D. Srivastava. 2022. "An IoT & AI-Assisted Framework for Agriculture Automation." In *2022 10th International Conference on Reliability, Infocom Technologies and Optimization (Trends and Future Directions) (ICRITO)*, Noida, India, 1–6. https://doi.org/10.1109/ICRITO56286.2022.9964567.
49. Khaki, S., L. Wang, and S. V. Archontoulis. 2020. "A CNN-RNN Framework for Crop Yield Prediction." *Frontiers in Plant Science* 10: 1750. https://doi.org/10.3389/fpls.2019.01750.
50. Majeed, Y., L. Fu, and L. He. 2024. "Editorial: Artificial Intelligence-of-Things (AIoT) in Precision Agriculture." *Frontiers in Plant Science* 15: 1369791. https://doi.org/10.3389/fpls.2024.1369791.

51. Saikai, Y., A. Peake, and K. Chenu. 2023. "Deep Reinforcement Learning for Irrigation Scheduling Using High-Dimensional Sensor Feedback." *PLOS Water* 2 (9): e0000169. https://doi.org/10.1371/journal.pwat.0000169.
52. Mim, M. I., F. Sultana, and M. R. Hasan. 2025. "AI-Powered Autonomous Farming: The Future of Sustainable Agriculture." *European Journal of Theoretical and Applied Sciences* 3 (1): 11–31. https://doi.org/10.59324/ejtas.2025.3(1).02.
53. Agrawal, J., and M. Y. Arafat. 2024. "Transforming Farming: A Review of AI-Powered UAV Technologies in Precision Agriculture." *Drones* 8 (11): 664. https://doi.org/10.3390/drones8110664.
54. Rajeev, P. N., and S. Joy. 2020. "CropIn: Addressing Farmer Poverty Through 'Connected' Farming." In Latha Poonamallee, Joanne Scillitoe, and Simy Joy (Eds.),*Socio-Tech Innovation*, 289–304. Cham: Springer International Publishing. https://doi.org/10.1007/978-3-030-39554-4_15.
55. Sharma, K., and S. K. Shivandu. 2024. "Integrating Artificial Intelligence and Internet of Things (IoT) for Enhanced Crop Monitoring and Management in Precision Agriculture." *Sensors International* 5: 100292. https://doi.org/10.1016/j.sintl.2024.100292.
56. Kamilaris, A., and F. X. Prenafeta-Boldú. 2018a. "A Review of the Use of Convolutional Neural Networks in Agriculture." *The Journal of Agricultural Science* 156 (3): 312–22. https://doi.org/10.1017/S0021859618000436.
57. Maes, W. H., and K. Steppe. 2019. "Perspectives for Remote Sensing with Unmanned Aerial Vehicles in Precision Agriculture." *Trends in Plant Science* 24 (2): 152–64. https://doi.org/10.1016/j.tplants.2018.11.007.
58. Tsouros, D. C., S. Bibi, and P. G. Sarigiannidis. 2019. "A Review on UAV-Based Applications for Precision Agriculture." *Information* 10 (11): 349. https://doi.org/10.3390/info10110349.
59. Kassem, H. S., B. A. Alotaibi, Y. A. Ghoneim, and A. M. Diab. 2021. "Mobile-Based Advisory Services for Sustainable Agriculture: Assessing Farmers' Information Behavior." *Information Development* 37 (3): 483–95. https://doi.org/10.1177/0266666920967979.
60. Fuentes-Peñailillo, F., K. Gutter, R. Vega, and G. C. Silva. 2024. "Transformative Technologies in Digital Agriculture: Leveraging Internet of Things, Remote Sensing, and Artificial Intelligence for Smart Crop Management." *Journal of Sensor and Actuator Networks* 13 (4): 39. https://doi.org/10.3390/jsan13040039.
61. Benti, N. E., M. D. Chaka, A. G. Semie, B. Warkineh, and T. Soromessa. 2024. "Transforming Agriculture with Machine Learning, Deep Learning, and IoT: Perspectives from Ethiopia—Challenges and Opportunities." *Discover Agriculture* 2 (1): 63. https://doi.org/10.1007/s44279-024-00066-7.
62. Kushwaha, Y. K., A. Joshi, R. K. Panigrahi, and A. Pandey. 2024. "Development of a Smart Irrigation Monitoring System Employing the Wireless Sensor Network for Agricultural Water Management." *Journal of Hydroinformatics* 26 (12): 3224–43. https://doi.org/10.2166/hydro.2024.241.
63. Lee, Y., S. Ha, X. Wang, S. Hahm, K. Lee, and J. Park. 2025. "An Automatic Irrigation System Based on Hourly Cumulative Evapotranspiration for Reducing Agricultural Water Usage." *Agriculture* 15 (3): 308. https://doi.org/10.3390/agriculture15030308.

64. Gautron, R., O.-A. Maillard, P. Preux, M. Corbeels, and R. Sabbadin. 2022. "Reinforcement Learning for Crop Management Support: Review, Prospects and Challenges." *Computers and Electronics in Agriculture* 200: 107182. https://doi.org/10.1016/j.compag.2022.107182.
65. Boopathi, S. 2024. "Sustainable Development Using IoT and AI Techniques for Water Utilization in Agriculture." In Rajeev Kumar, Abu Bakar Abdul Hamid, Dato' Dr Noor Inayah Binti Binti Ya'akub, Hari Om Sharan, and Sandeep Kumar (Eds.), *Sustainable Development in AI, Blockchain, and E-Governance Applications*, 204–28. IGI Global Scientific Publishing. https://doi.org/10.4018/979-8-3693-1722-8.ch012.
66. Guebsi, R., S. Mami, and K. Chokmani. 2024. "Drones in Precision Agriculture: A Comprehensive Review of Applications, Technologies, and Challenges." *Drones* 8 (11): 686. https://doi.org/10.3390/drones8110686.
67. Li, L., S. Zhang, and B. Wang. 2021. "Plant Disease Detection and Classification by Deep Learning — A Review." *IEEE Access* 9: 56683–98. https://doi.org/10.1109/ACCESS.2021.3069646.
68. Kuradusenge, M., E. Hitimana, K. Mtonga, A. Gatera, J. Habiyaremye, J. Ngabonziza, D. Hanyurwimfura, P. Rukundo, and A. Mukasine. 2024. "SMART-CYPS: An Intelligent Internet of Things and Machine Learning Powered Crop Yield Prediction System for Food Security." *Discover Internet of Things* 4 (1): 23. https://doi.org/10.1007/s43926-024-00079-0.
69. Amiri-Zarandi, M., M. Hazrati Fard, S. Yousefinaghani, M. Kaviani, and R. Dara. 2022. "A Platform Approach to Smart Farm Information Processing." *Agriculture* 12 (6): 838. https://doi.org/10.3390/agriculture12060838.
70. Amiri-Zarandi, M., R. A. Dara, E. Duncan, and E. D. G. Fraser. 2022. "Big Data Privacy in Smart Farming: A Review." *Sustainability* 14 (15): 9120. https://doi.org/10.3390/su14159120.
71. Khanna, M., S. S. Atallah, T. Heckelei, L. Wu, and H. Storm. 2024. "Economics of the Adoption of Artificial Intelligence–Based Digital Technologies in Agriculture." *Annual Review of Resource Economics* 16 (1): 41–61. https://doi.org/10.1146/annurev-resource-101623-092515.
72. Farooq, M. S., S. Riaz, A. Abid, T. Umer, and Y. Bin Zikria. 2020. "Role of IoT Technology in Agriculture: A Systematic Literature Review." *Electronics* 9 (2): 319. https://doi.org/10.3390/electronics9020319.
73. Reddy, K. V., B. S. K. Reddy, V. Goutham, M. Mahesh, J. S. Nisha, G. Palanisamy, M. Golla, S. Purushothaman, K. R. Reddy, and V. Ramkumar. 2024. "Edge AI in Sustainable Farming: Deep Learning-Driven IoT Framework to Safeguard Crops from Wildlife Threats." *IEEE Access* 12: 77707–23. https://doi.org/10.1109/ACCESS.2024.3406585.
74. Kootstra, G., X. Wang, P. M. Blok, S. Hemming, and E. van Henten. 2021. "Selective Harvesting Robotics: Current Research, Trends, and Future Directions." *Current Robotics Reports* 2 (1): 95–104. https://doi.org/10.1007/s43154-020-00034-1.
75. Elhattab, K., and S. Elatar. 2024. "Survey of IoT and AI Applications: Future Challenges and Opportunities in Agriculture." *Indonesian Journal of Electrical Engineering and Computer Science* 36 (3): 1655–63. https://doi.org/10.11591/ijeecs.v36.i3.
76. Editorial. 2024. Feeding the Future Global Population. *Nature Communications* 15 (1): 222. https://doi.org/10.1038/s41467-023-44588-y.

77. Vallejo-Gómez, D., M. Osorio, and C. A. Hincapié. 2023. "Smart Irrigation Systems in Agriculture: A Systematic Review." *Agronomy* 13 (2): 342. https://doi.org/10.3390/agronomy13020342.
78. Ansari, M. A., N. Ravisankar, M. Shamim, M. Rani, A. K. Prusty, R. Singh, R. K. J., H. Joshi, S. Kumar, A. S. Panwar, and M. Kumar. 2024. "Accounting of Carbon Sequestration and Tradeoff under Various Climatic Scenarios in Alternative Agricultural System: A Comprehensive Framework toward Carbon Neutrality." *Frontiers in Sustainable Food Systems* 8: 1371255. https://doi.org/10.3389/fsufs.2024.1371255.
79. MacPherson, J., A. Voglhuber-Slavinsky, M. Olbrisch, P. Schöbel, E. Dönitz, I. Mouratiadou, and K. Helming. 2022. "Future Agricultural Systems and the Role of Digitalization for Achieving Sustainability Goals: A Review." *Agronomy for Sustainable Development* 42 (4): 70. https://doi.org/10.1007/s13593-022-00792-6.
80. Kumar, A., N. Karn, and H. Sharma. 2024. "IoT, AI, and Robotics Applications in the Agriculture Sector." In Suchismita Satapathy and Kamalakanta Muduli (Eds.), *IoT, AI, and Robotics Applications in Agriculture*, 243–272. https://doi.org/10.4018/979-8-3693-3583-3.ch014.
81. Kulkarni, S. D. 2025. "Sustainable Precision Agriculture: Integrating Artificial Intelligence and IoT for Optimized Green Farming Practices." *Journal of Informatics Education and Research* 5 (1): 312–322. https://doi.org/10.52783/jier.v5i1.2000.

Chapter 11

Predictive analytics for crop yield optimization

Using AI to forecast and maximize agricultural output

B. Mohan Rao and Aman Kumar

11.1 INTRODUCTION

The global agricultural sector faces mounting pressure to meet the demands of a rapidly growing population, expected to exceed 9.7 billion by 2050 [1]. Traditional farming methods, reliant on manual labor and intuition, are increasingly insufficient to tackle challenges such as climate change, resource scarcity, and pest outbreaks. To address these issues, the integration of predictive analytics and artificial intelligence (AI) has emerged as a transformative solution for optimizing crop yields and enhancing agricultural sustainability [2].

Predictive analytics leverages historical and real-time data from various sources, including Internet of Things (IoT)-enabled sensors, satellite imagery, and climate records, to forecast agricultural outcomes. These technologies enable farmers to make informed decisions about plant schedules, resource allocation, and risk management [3]. For instance, IoT sensors embedded in fields monitor soil moisture, temperature, and nutrient levels, transmitting data to AI systems for analysis and actionable insights [4].

Machine learning models such as regression algorithms, decision trees, and deep learning frameworks are pivotal in these applications. These models analyze complex datasets to predict crop yields, identify optimal planting strategies, and detect early signs of disease [5]. Furthermore, the synergy between AI and IoT has facilitated the development of precision farming techniques, ensuring that resources like water and fertilizers are used efficiently, minimizing waste while maximizing output [6].

The adoption of predictive analytics in agriculture is not without challenges. Issues such as data variability, high costs of technology implementation, and limited digital literacy among farmers need to be addressed. However, as demonstrated by successful projects in rice farming in Asia and vineyard monitoring in Europe, these technologies have the potential to revolutionize agriculture on a global scale [7,8].

This chapter explores the role of predictive analytics in crop yield optimization, focusing on its applications, challenges, and future directions. Real-world

DOI: 10.1201/9781003589730-11

case studies illustrate how AI-driven solutions are transforming farming practices, enabling sustainable agriculture and food security (Figure 11.1).

11.1.1 Role of data in crop yield prediction

The foundation of predictive analytics in agriculture is data is a critical asset for any AI-driven application. In this context, various data sources contribute to enhancing the accuracy and relevance of crop yield forecasts. These include IoT-enabled sensors, satellite imagery, and historical weather data. IoT sensors collect data on soil moisture, temperature, and nutrient levels, which provide real-time insights for timely interventions. Satellite imagery, on the other hand, offers a bird's-eye view of crop health, detecting early signs of disease or stress that may not be visible to the naked eye [3]. Additionally, historical weather data, such as temperature and rainfall patterns, are essential for building predictive models that account for environmental variability [4] as shown in Figure 11.2.

Figure 11.1 IoT sensors in a farmland setup, demonstrating real-time data collection.

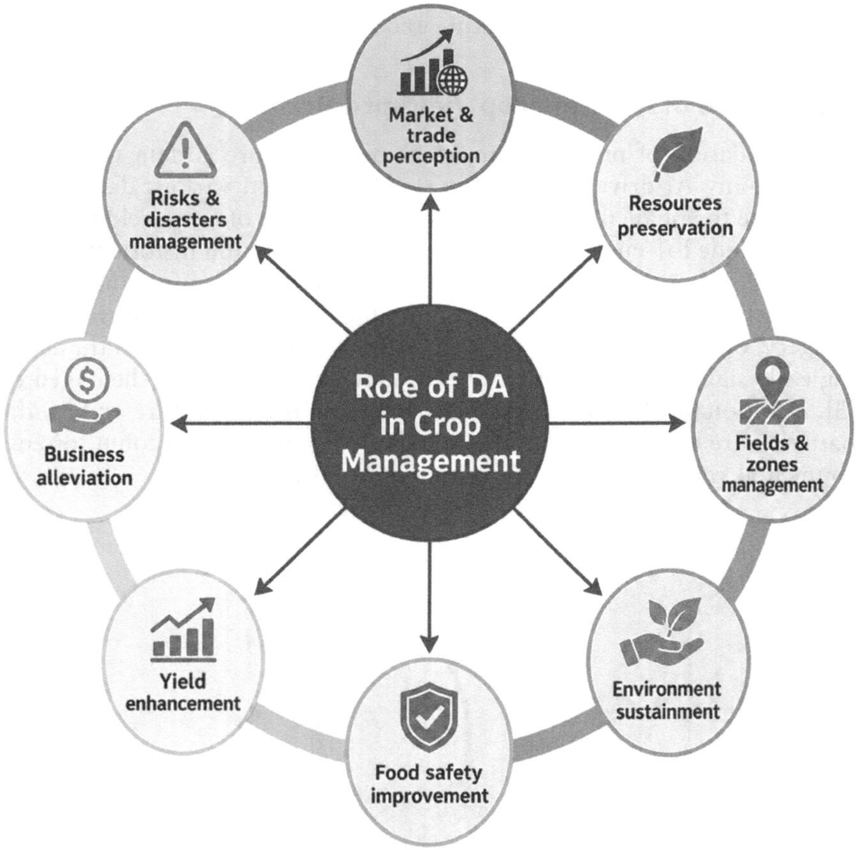

Figure 11.2 Role of data analytics in crop management.

Challenges in data collection include ensuring consistent quality and avoiding gaps in data from rural areas, where access to technology may be limited. Despite these challenges, combining multiple data sources allows for a more comprehensive understanding of the factors influencing crop yields [5].

11.2 AI TECHNIQUES FOR CROP YIELD PREDICTION

11.2.1 Regression models

Regression models are foundational techniques in predictive analytics. Linear regression, polynomial regression, and other forms of regression are widely used to predict crop yields based on various environmental factors like temperature, precipitation, and soil conditions. In linear regression, the relationship between the dependent variable (crop yield) and the independent variables (weather parameters, soil moisture, etc.) is modeled using a straight line. Polynomial regression is an extension that models the relationship with higher-degree polynomials, providing better accuracy when the relationship is non-linear [6].

11.2.2 Decision trees

Decision trees are another AI technique commonly used in crop yield prediction. These models partition the data into subsets based on different features and make predictions by traversing from the root node to leaf nodes. For instance, a decision tree model might predict crop yield by considering features like soil type, irrigation method, and climate factors. It is particularly valuable for handling complex and categorical data, making it an effective tool for farming applications [9]. The decision tree's simplicity and transparency make it easy to interpret and understand, a crucial aspect for farmers who need actionable insights. Decision trees can also be enhanced by ensemble learning methods like Random Forests, which combine multiple decision trees to improve prediction accuracy [10].

11.2.3 Neural networks and deep learning

Artificial neural networks (ANNs) and deep learning models, particularly Convolutional Neural Networks (CNNs), are revolutionizing the field of crop yield prediction, especially when dealing with complex, high-dimensional data. CNNs are particularly well-suited for processing satellite imagery and remote sensing data to assess crop health and predict yield. CNNs learn features from raw data without manual feature extraction, enabling accurate identification of patterns in images, such as identifying areas of crop stress due to pests or drought [5]. Moreover, Recurrent Neural Networks (RNNs), particularly Long Short-Term Memory (LSTM) networks, are effective for time-series data analysis, such as weather patterns over multiple seasons. LSTMs are adept at learning and remembering long-term dependencies in data, which makes them ideal for predicting crop yields in response to changing weather and climatic conditions over time [11].

11.2.4 Ensemble learning

Ensemble learning combines the strengths of multiple machine learning models to improve prediction performance. One of the most popular ensemble learning techniques used for crop yield prediction is Random Forest, which builds a collection of decision trees, each trained on a random subset of the data. By averaging the outputs of all the trees, the model reduces variance and prevents overfitting, leading to more robust predictions [12]. Another ensemble technique is Gradient Boosting, which sequentially builds trees, with each new tree correcting the errors of the previous ones. This approach has been widely adopted for crop yield forecasting as it can efficiently handle large datasets and produce highly accurate predictions [13].

11.2.5 Support vector machines

Support Vector Machines (SVMs) are supervised learning models that classify data by finding the hyperplane that best separates different classes. For crop yield prediction, SVM can be adapted for regression tasks to predict the amount of yield based on input features like soil fertility, weather conditions, and crop type. SVM models are known for their ability to handle both linear and non-linear relationships between features, making them suitable for complex agricultural data [14].

11.2.6 Fuzzy logic

Fuzzy logic is a technique used to handle uncertainty in data, which is a common problem in agriculture due to the unpredictable nature of weather and climate. Fuzzy systems use membership functions to represent data in degrees of truth rather than absolute values. For example, instead of classifying a soil's moisture level as either "dry" or "wet," fuzzy logic allows it to be represented as a percentage, such as 70% wet or 30% dry, depending on the data available. This flexibility makes fuzzy logic a valuable tool for crop yield prediction under uncertain conditions [15].

11.2.7 Clustering and classification

Clustering and classification algorithms, such as K-means clustering and *K*-Nearest Neighbors (KNNs), are widely used to categorize regions of land based on soil properties, irrigation techniques, and crop types. These algorithms group similar data points together, helping to identify patterns and predict yields based on the features of a particular region. For instance, clustering techniques can group fields with similar soil types and weather conditions, allowing farmers to optimize their strategies for planting and irrigation [16].

11.2.8 Predictive modeling with time-series data

Time-series analysis is crucial for predicting crop yields over time, particularly when considering factors like weather patterns, rainfall, and temperature fluctuations. ARIMA (AutoRegressive Integrated Moving Average) models are commonly used for time-series forecasting in agriculture. These models analyze historical data to predict future crop yields based on expected environmental conditions. Similarly, LSTM networks, a type of RNN, have become highly effective in analyzing time-series data and making long-term predictions based on trends observed in past weather and crop yield data [17].

11.2.9 Integration of AI and IoT for real-time data processing

AI-powered models are increasingly being integrated with IoT systems to enable real-time data processing in agriculture. IoT devices such as soil sensors, weather stations, and drones collect valuable data, which is then processed by AI algorithms to provide real-time insights. This integration allows farmers to make instant decisions, such as adjusting irrigation systems or applying fertilizers only when necessary, leading to higher crop yields and resource savings. The synergy between AI and IoT has given rise to the concept of smart farming, where automated systems manage and optimize crop production [18].

11.2.10 Edge computing in AI for agriculture

Edge computing is another emerging technology that is transforming AI applications in agriculture. Instead of sending all collected data to a centralized cloud server for processing, edge computing allows data to be processed locally, at the "edge" of the network, near the source of data collection. This reduces latency and ensures quicker decision-making. For example, edge devices can process sensor data from farms in real time, and AI models can then make predictions on-site, helping farmers make immediate adjustments to farming operations like irrigation or pesticide spraying [19].

11.3 APPLICATIONS OF PREDICTIVE ANALYTICS IN AGRICULTURE

Predictive analytics plays a crucial role in optimizing crop yields by leveraging AI and machine learning techniques to forecast outcomes based on historical data, weather patterns, soil conditions, and other variables. These applications not only increase agricultural efficiency but also enhance decision-making in real time, enabling farmers to take proactive steps to ensure optimal production. Below are key applications of predictive analytics in crop yield optimization:

11.3.1 Weather prediction and crop planning

One of the most significant applications of predictive analytics in agriculture is its ability to forecast weather conditions and their potential impact on crop production. Using historical weather data, AI models can predict future weather patterns such as rainfall, temperature fluctuations, and droughts. These predictions enable farmers to make data-driven decisions regarding planting schedules, irrigation requirements, and crop selection. By accurately forecasting weather conditions, predictive analytics helps farmers reduce risks associated with unpredictable weather, leading to better crop management and increased yield [20].

11.3.2 Soil health monitoring and fertilizer optimization

Predictive analytics is increasingly being used for soil health monitoring, which is essential for maximizing crop yield. By analyzing historical soil data and real-time sensor readings (e.g., soil moisture, pH levels, and nutrient content), AI models can predict soil degradation or nutrient imbalances. These insights allow farmers to optimize the application of fertilizers, water, and other soil treatments. The optimization of fertilizer usage not only improves crop productivity but also reduces environmental harm by minimizing excess fertilizer runoff, a major issue in sustainable farming practices [21].

11.3.3 Pest and disease forecasting

AI-powered predictive models are also being applied to forecast pest and disease outbreaks. These models analyze data from various sources such as sensor networks, satellite imagery, and field reports to predict the likelihood of pest infestations or crop diseases. Early detection of pests and diseases enables farmers to take preventive measures before the problems escalate, thereby saving crops and reducing the need for harmful pesticides. This application of predictive analytics contributes to healthier crops and supports sustainable agriculture by minimizing pesticide usage [22].

11.3.4 Precision irrigation systems

Predictive analytics is at the heart of precision irrigation systems that aim to optimize water usage. By integrating weather forecasts, soil moisture levels, and crop water requirements, AI models can predict the optimal amount of water needed for irrigation. These models help in reducing water wastage and ensuring that crops receive the precise amount of water required at each growth stage. This application is particularly important in regions facing water scarcity and helps farmers achieve high yields while conserving valuable water resources [23].

11.3.5 Yield prediction for supply chain optimization

Predictive analytics also aids in crop yield prediction, which is crucial for supply chain optimization. By forecasting the expected crop yields, farmers, distributors, and retailers can plan their activities, reducing waste and ensuring a steady supply of produce, as shown in Figure 11.3. This also helps in optimizing inventory management and pricing strategies. Accurate yield predictions can also assist in better matching supply with demand, particularly in perishable crops, thereby minimizing market inefficiencies and ensuring that food reaches consumers at the right time [24].

11.3.6 AI-driven decision support systems for farmers

AI-driven decision support systems (DSS) have gained significant attention in agriculture due to their ability to provide actionable insights for farmers. These systems integrate multiple data sources, including weather forecasts,

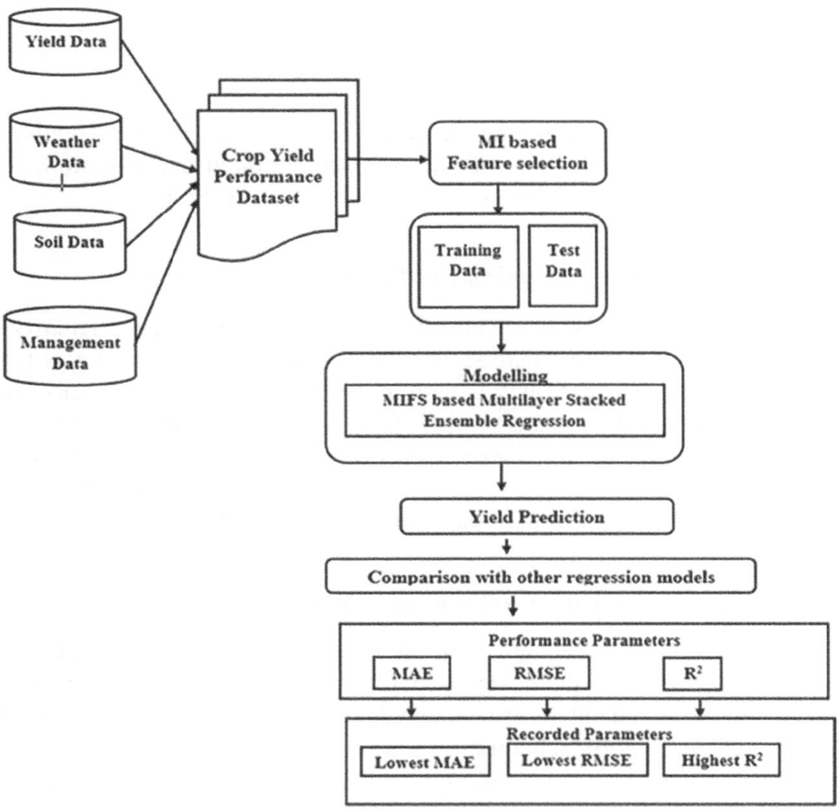

Figure 11.3 Crop yield prediction.

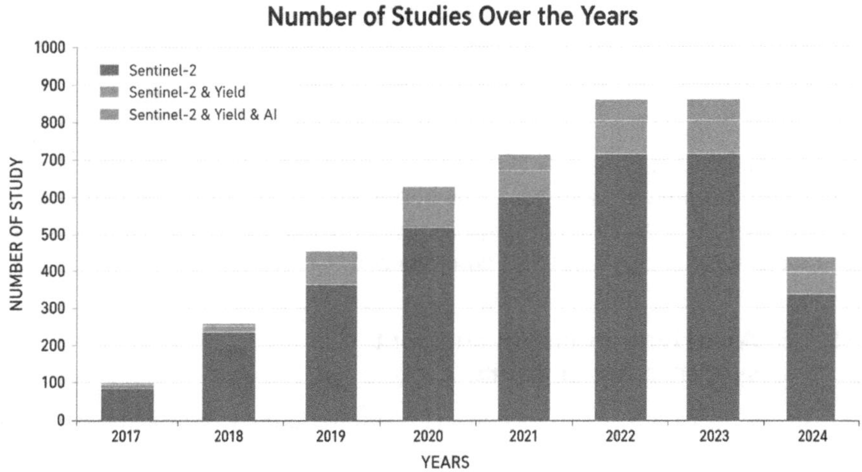

Figure 11.4 Artificial intelligence techniques in crop yield estimation based on sentinel-2 data.

soil conditions, crop health monitoring, and real-time market trends, to assist farmers in making informed decisions about planting, irrigation, pest control, and harvesting. The data are shown in Figure 11.4.

11.3.7 Predictive analytics for farm management

Predictive analytics plays a crucial role in farm management, particularly in crop planning and resource allocation. Using historical data and real-time environmental inputs, AI-based systems forecast crop performance and provide advice on the optimal timing for sowing, irrigation, and fertilization. For example, machine learning algorithms can predict when a crop is likely to be ready for harvest based on growth patterns and climatic conditions, allowing farmers to plan labor and resources effectively [25].

11.3.8 Optimization of irrigation systems

Water scarcity is a significant issue in agriculture, making the optimization of irrigation systems a key area for AI implementation. AI models, such as those based on reinforcement learning and multi-agent systems, can optimize irrigation schedules based on real-time data from soil moisture sensors, weather forecasts, and crop water needs. These models can continuously learn and adapt, ensuring that water is used efficiently while avoiding under- or over-irrigation [26].

For example, reinforcement learning algorithms can be used to control irrigation systems by rewarding the system for achieving optimal water usage while penalizing overuse. By incorporating data from various IoT devices and historical weather patterns, these AI systems can provide real-time recommendations on the exact amount of water to be applied, significantly improving water efficiency [27].

11.3.9 Pest and disease detection

AI-powered systems are also playing a crucial role in pest and disease detection. By analyzing images captured through drones, satellites, and smartphones, deep learning models can identify signs of pest infestations or plant diseases at early stages, as shown in Figure 11.5. These systems use CNNs to analyze the images and classify the type of pest or disease, allowing farmers to take immediate action before the problem escalates [28].

Additionally, AI-based DSS systems integrate data from multiple sensors to predict the likelihood of pest outbreaks or diseases based on environmental conditions, crop species, and historical data. This allows farmers to apply pesticides or treatments only when necessary, reducing costs and minimizing environmental impact [29].

11.3.10 Market forecasting and price prediction

AI is also transforming the way farmers approach market forecasting and price prediction. Using historical data, market trends, and factors like weather conditions and crop yield, AI models can predict future crop prices and demand. These predictions help farmers make informed decisions about what crops to grow and when to sell them, thereby optimizing profits and reducing financial risk [30]. Machine learning algorithms such as SVMs and Random Forests are particularly useful for predicting commodity prices by analyzing a variety of economic indicators, market reports, and supply-demand patterns. By forecasting market prices accurately, these AI-based systems assist farmers in planning their production cycles and managing their supply chain effectively [31].

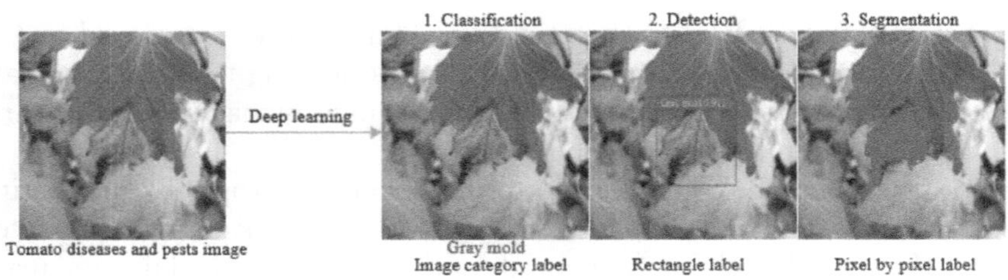

Figure 11.5 Pest and disease detection.

11.3.11 Climate adaptation and resilience

AI-based DSS systems are crucial for enhancing the resilience of agriculture to climate change. With climate change causing increasingly erratic weather patterns, farmers need tools that can help them adapt to these changes. AI models can predict the likelihood of extreme weather events such as droughts, floods, or heat waves and recommend crop varieties that are more resilient to these conditions [32].

AI systems can analyze climate data and recommend the most suitable crops for different weather scenarios. Moreover, AI can simulate various farming strategies under changing climate conditions, providing insights into how to adjust planting schedules, irrigation methods, and crop protection strategies to mitigate climate risks [33].

11.3.12 Real-time monitoring of crop health

AI systems, integrated with IoT devices, allow for real-time monitoring of crop health. Using data from sensors that measure environmental conditions, as well as images from drones and satellites, AI systems can detect early signs of crop stress or disease, providing farmers with the opportunity to intervene before the situation worsens. This continuous monitoring is crucial for optimizing crop yield and reducing losses due to undetected issues [34].

AI-driven systems utilize advanced image recognition techniques and sensor fusion to monitor crop growth stages, nutrient levels, and potential stress factors such as pest attacks or weather extremes. This provides farmers with detailed insights into the health of their crops, enabling timely interventions and better resource management [35].

11.4 REAL-WORLD CASE STUDIES IN CROP YIELD OPTIMIZATION

Several global case studies highlight the successful use of predictive analytics for crop yield optimization, as shown in Figure 11.6.

1. Case Study 1: Rice Farming in Asia: AI models were deployed to predict rice yield in several Asian countries, leading to a 15% increase in crop production through better resource management and early intervention for pests [36].
2. Case Study 2: Wheat Yield Optimization in Europe: By integrating satellite data and weather forecasts, AI models accurately predicted wheat yields, helping farmers adjust their planting schedules and manage water resources more effectively [37].

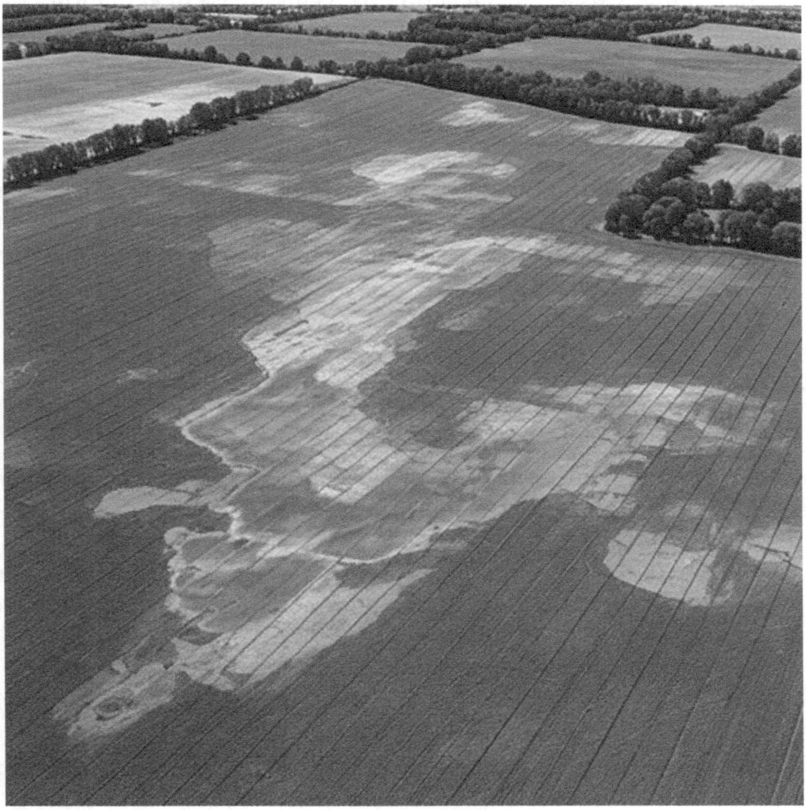

Figure 11.6 Satellite image of a field with highlighted crop health zones, generated using AI.

3. Case Study 3: Vineyard Management in Europe: AI-based systems optimized grape quality by predicting harvest times and managing water usage based on real-time environmental data, leading to improved wine quality and yield [38].

11.5 CHALLENGES AND LIMITATIONS OF PREDICTIVE ANALYTICS IN AGRICULTURE

Despite the promising potential of predictive analytics, several challenges hinder its widespread adoption:

1. Data Quality and Availability: Inaccurate or sparse data, especially in rural areas, can affect model accuracy. Ensuring high-quality, consistent data is crucial for reliable predictions [39].

2. High Initial Costs: Implementing AI technologies in agriculture often involves significant upfront investments, including the cost of IoT devices, software, and AI infrastructure [40].
3. Technical Literacy: Many farmers, particularly in developing regions, may lack the technical expertise to use advanced AI tools effectively. This barrier requires training and capacity-building efforts [41].
4. Integration with Existing Systems: Many farmers use traditional methods for crop management, and integrating AI systems with these existing practices may be challenging due to technological resistance or lack of infrastructure [42].

11.6 FUTURE DIRECTIONS IN PREDICTIVE ANALYTICS FOR AGRICULTURE

The future of predictive analytics in agriculture holds immense promise. Innovations are expected in the following areas:

1. Blockchain for Data Sharing: Block chain can be used to ensure the security and transparency of agricultural data, allowing for seamless sharing between stakeholders [43].
2. AI-Driven Drones for Real-Time Monitoring: Drones equipped with AI systems can provide real-time insights into crop health and growth, enabling faster decision-making [44].
3. Climate Adaptation Strategies: Predictive models will increasingly incorporate climate adaptation strategies, helping farmers adapt to the effects of climate change, such as erratic rainfall and temperature shifts [45].
4. Affordable AI Solutions for Small-Scale Farmers: With advances in edge computing and low-cost sensors, AI solutions will become more affordable and accessible to smallholder farmers worldwide [46].

REFERENCES

1. United Nations, *World Population Prospects*, 2022.
2. Tripathi, Arpit, Harshlata Vishwakarma, Rajneesh Kumar Patel, Anju Shukla, and Siddharth Singh Chouhan. AI-enhanced agriculture: Cultivating the future. In *AgriTech Revolution: Next-Gen Solutions and Challenges in Modern Agriculture*, pp. 1–14. Singapore: Springer Nature Singapore, 2025.
3. Collins, L. M., and L. M. Smith. Smart agri-systems for the pig industry. *Animal* 16 (2022): 100518.
4. Victor, Brandon, Aiden Nibali, and Zhen He. A systematic review of the use of deep learning in satellite imagery for agriculture. *IEEE Journal of Selected Topics in Applied Earth Observations and Remote Sensing* 18 (2024): 2297–2316.

5. Ganatra, Nilay, and Atul Patel. Deep learning methods and applications for precision agriculture. *Machine Learning for Predictive Analysis: Proceedings of ICTIS 2020* (2020): 515–527.
6. Zaman, Qamar U. Precision agriculture technology: A pathway toward sustainable agriculture. In *Precision agriculture*, pp. 1–17. Academic Press, 2023.
7. Ansari, Andrianto, Yu-Pin Lin, and Huu-Sheng Lur. Evaluating and adapting climate change impacts on rice production in Indonesia: A case study of the Keduang subwatershed, Central Java. *Environments* 8, no. 11 (2021): 117.
8. Cividino, Sirio RS, Alessio Cappelli, Paolo Belluco, Fabiano Rinaldi, Lena Avramovic, and Mauro Zaninelli. Preliminary study of vision-based artificial intelligence application to evaluate occupational risks in viticulture. *Sensors* 25, no. 21 (2025): 6749.
9. Kim, Nari, Sang-Il Na, Chan-Won Park, Morang Huh, Jaiho Oh, Kyung-Ja Ha, Jaeil Cho, and Yang-Won Lee. An artificial intelligence approach to prediction of corn yields under extreme weather conditions using satellite and meteorological data. *Applied Sciences* 10, no. 11 (2020): 3785.
10. Zhang, Y., *AI and IoT in Precision Farming*, AgriTech Review, 2022.
11. Kamilaris, Andreas, and Francesc X. Prenafeta-Boldú. Deep learning in agriculture: A survey. *Computers and Electronics in Agriculture* 147 (2018): 70–90.
12. Choi, Ji Won, Mohamad Soleh Hidayat, Soo Been Cho, Woon-Ha Hwang, Hoonsoo Lee, Byoung-Kwan Cho, Moon S. Kim, Insuck Baek, and Geonwoo Kim. Recent trends in machine learning, deep learning, ensemble learning, and explainable artificial intelligence techniques for evaluating crop yields under abnormal climate conditions. *Plants* 14, no. 18 (2025): 2841.
13. Kumar, R., *Gradient Boosting and Its Applications in Agriculture*, AgriTech Innovations, 2022.
14. Kok, Zhi Hong, Abdul Rashid Mohamed Shariff, Meftah Salem M. Alfatni, and Siti Khairunniza-Bejo. Support vector machine in precision agriculture: A review. *Computers and Electronics in Agriculture* 191 (2021): 106546.
15. Erdoğdu, Aylin, Faruk Dayi, Ferah Yildiz, Ahmet Yanik, and Farshad Ganji. "Combining fuzzy logic and genetic algorithms to optimize cost, time and quality in modern agriculture." *Sustainability* 17, no. 7 (2025): 2829.
16. Gavioli, Alan, Eduardo Godoy de Souza, Claudio Leones Bazzi, Kelyn Schenatto, and Nelson Miguel Betzek. Identification of management zones in precision agriculture: An evaluation of alternative cluster analysis methods. *Biosystems Engineering* 181 (2019): 86–102.
17. Pandit, Pramit, Atish Sagar, Bikramjeet Ghose, Prithwiraj Dey, Moumita Paul, Saeed Alqadhi, Javed Mallick, Hussein Almohamad, and Hazem Ghassan Abdo. Hybrid time series models with exogenous variable for improved yield forecasting of major Rabi crops in India. *Scientific Reports* 13, no. 1 (2023): 22240.
18. Alazzai, Waleed K., Baydaa Sh Z. Abood, Hassan M. Al-Jawahry, and Mohammed Kadhim Obaid. Precision farming: the power of AI and IoT technologies. In *E3S web of conferences*, vol. 491, p. 04006. EDP Sciences, 2024.
19. El Jarroudi, Moussa, Louis Kouadio, Philippe Delfosse, Clive H. Bock, Anne-Katrin Mahlein, Xavier Fettweis, Benoit Mercatoris, Frank Adams, Jillian M. Lenné, and Said Hamdioui. Leveraging edge artificial intelligence for sustainable agriculture. *Nature Sustainability* 7, no. 7 (2024): 846–854.

20. Sridhara, Shankarappa, Konapura Nagaraja Manoj, Pradeep Gopakkali, Girish R. Kashyap, Bappa Das, Kamlesh Kumar Singh, and Amit Kumar Srivastava. Evaluation of machine learning approaches for prediction of pigeon pea yield based on weather parameters in India. *International Journal of Biometeorology* 67, no. 1 (2023): 165–180.
21. Swain, Smita Rani, and Rakesh Kumar Gupta. Smart water, soil, and environment management system: An approach toward sustainable farming. In *Smart Horticulture Production*, pp. 239–252. Apple Academic Press, 2026.
22. Vidya Madhuri, E., J. S. Rupali, S. P. Sharan, N. Sai Pooja, G. S. Sujatha, Dhananjaya P. Singh, Khurshid Ahmad, Amrender Kumar, and Ratna Prabha. Transforming pest management with artificial intelligence technologies: The future of crop protection. *Journal of Crop Health* 77, no. 2 (2025): 48.
23. Shenoy, Harish, Vijay Laxmi, and Huchchappa Jamakhandi. Precision irrigation technologies in the era of climate stress: Methods, performance, and water use efficiency: A systematic review. *Management* 8: 9.
24. Kumar, Sandeep, and Prabir Sarkar. Machine learning–enabled decision support for planning and control in sustainable agricultural supply chains. *Production Planning & Control* (2026): 1–27.
25. Patel, Govind, and Manish Patel. Agropredml: Crop yield prediction in smart agriculture using machine learning. *International Journal of Advances in Signal and Image Sciences* (2025): 475–486.
26. Mohammed, Naz, Hussain Thahir Tahir, Skala Salam Othman, Dalya Aydeen Hawar, Murad Abdullah Abdulkadir, and Cem Korkmaz. Smart irrigation systems: A comprehensive review of IoT, AI, and sustainable agriculture technologies. A review article. *Kirkuk University Journal for Agricultural Sciences (KUJAS)* 16, no. 4 (2025): 266–281.
27. Kumar, Sandeep, and Prabir Sarkar. Machine learning–enabled decision support for planning and control in sustainable agricultural supply chains. *Production Planning & Control* (2026): 1–27.
28. Zhu, Hongyan, Chengzhi Lin, Gengqi Liu, Dani Wang, Shuai Qin, Anjie Li, Jun-Li Xu, and Yong He. Intelligent agriculture: Deep learning in UAV-based remote sensing imagery for crop diseases and pests detection. *Frontiers in Plant Science* 15 (2024): 1435016.
29. Smith, T., *Predictive Analytics for Pest Management in Agriculture*, AI in Crop Protection, 2021.
30. Anderson, C., *AI for Predicting Crop Prices and Market Trends*, AgriTech Review, 2021.
31. Sharma, Priyanka, Pankaj Dadheech, Nagender Aneja, and Sandhya Aneja. Predicting agriculture yields based on machine learning using regression and deep learning. *IEEE Access* 11 (2023): 111255–111264.
32. Mishra, Vidushi, Nandini Chaurasia, Divyansh Bajpai, Manoj Kumar Mishra, Pankaj Gupta, and Sanjay Mishra. AI Integrated Biotechnology versus Climate Resilient Agriculture, Healthcare and Food System: An Overview. *IOSR Journal of Biotechnology and Biochemistry (IOSR-JBB)*. ISSN: 2455-264X, Volume 11, Issue 5 Series 1 (Sept. – Oct. 2025), pp. 45–52. www.iosrjournals.org.
33. Kumar, K. Vimal, and Ghazaala Yasmin. Integrating climate data for agricultural resilience using geospatial approaches. In Agricultural Insights from Space, pp. 151–166. Academic Press, 2026.

34. Chen, Liang-Bi, Guan-Zhi Huang, Xiang-Rui Huang, and Wei-Chien Wang. A self-supervised learning-based intelligent greenhouse orchid growth inspection system for precision agriculture. *IEEE Sensors Journal* 22, no. 24 (2022): 24567–24577.
35. Martin, R. John, Ruchi Mittal, Varun Malik, Fathe Jeribi, Shams Tabrez Siddiqui, Mohammad Alamgir Hossain, and S. L. Swapna. XAI-powered smart agriculture framework for enhancing food productivity and sustainability. *IEEE Access* 12 (2024): 168412–168427.
36. Srivastava, Shobhit, Shikha Singh, Sujai Tripathi, and Shrawan Kumar Pandey. AI-powered smart agriculture advisor. *2025 3rd International Conference on IoT, Communication and Automation Technology (ICICAT)*, pp. 1–6. IEEE, 2025.
37. Vitale, Paolo, Jose Crossa, Abelardo Montesinos-López, Karim Ammar, Susanne Dreisigacker, Velu Govindan, Osval Montesinos-López, et al. Multi-omics prediction for yellow rust in bread and durum wheat through conventional and Ai-based frameworks. *Scientific Reports* (2026).
38. Rafique, Rizwan, Touqeer Ahmad, Muhammad Azam Khan, Mukhtar Ahmed, and Gerrit Hoogenboom. Developing a simple and efficient modeling solution for predicting key phenological stages of table grapes in a non-traditional viticulture zone in South Asia. *International Journal of Biometeorology* 68, no. 8 (2024): 1587–1601.
39. Patel, Chandra Shekhar, Krishna Kumar Verma, and Tryambak Hiwarkar. Use of data mining techniques in agriculture. *International Journal of Advanced Research in Computer Science* 17, no. 2 (2026): 40.
40. Smith, Matthew J. Getting value from artificial intelligence in agriculture. *Animal Production Science* 60, no. 1 (2019): 46–54.
41. Huang, Ke, Caoxin Chen, Hongyu Wu, Yi Su, Xiaoting Wu, Bo Huang, and Jiangjun Wan. A study on nonlinear characteristics and interaction effects in farmers' adoption of agricultural technologies based on an improved technology acceptance model and explainable artificial intelligence. *Agriculture* 16, no. 6 (2026): 678.
42. Ali, Zulfiqar, Asif Muhammad, Nangkyeong Lee, Muhammad Waqar, and Seung Won Lee. Artificial intelligence for sustainable agriculture: A comprehensive review of AI-driven technologies in crop production. *Sustainability* 17, no. 5 (2025): 2281.
43. Yap, Teck Lee, Rajkishore Nayak, Nhung TH Vu, Duy Tung Bui, Thi Thu Tra Pham, and Darcy WE Allen. Adopting blockchain-based traceability in the fruit supply chain in a developing economy: Facilitators and barriers. *Information Technology & People* 38, no. 1 (2025): 419–441.
44. Phang, Swee King, Tsai Hou Adam Chiang, Ari Happonen, and Miko May Lee Chang. From satellite to UAV-based remote sensing: A review on precision agriculture. *IEEE Access* 11 (2023): 127057–127076.
45. Singh, Sudhanshu, K. Srikanth Reddy, Malay K. Bhowmick, Ashish Kumar Srivastava, Sunil Kumar, and Panneerselvam Peramaiyan. Accelerating climate adaptation with big data analytics and ICTs. *Advances in Agri-Food Systems: Volume I*, pp. 179–196. Singapore: Springer Nature Singapore, 2025.
46. Das, Suman Kumar, and Pujyasmita Nayak. Integration of IoT-AI powered local weather forecasting: A game-changer for agriculture. *arXiv preprint arXiv:2501.14754* (2024).

Chapter 12

IoT-driven precision agriculture

Connected farming for a sustainable future

Vikrant Verma, Vipin Prakash Yadav, and Dipti Y. Sakhare

12.1 INTRODUCTION

Precision agriculture is an unconventional and progressive farming approach leveraging the cutting-edge technologies to optimize agricultural practices, enhancing productivity with efficient and effective utilization of inputs and ubiquitous connectivity. It plays a decisive role in bringing out sustainability in farming through appropriate sensing, communication and performing corrective actions to reduce wastage of resources [1]. Precision agriculture is revolutionizing farming by integrating technologies such as Internet of Things (IoT) to deploy systems for smart irrigation, smart fertilization, smart weed and pesticide management based on real-time data for soil conditions, crop conditions and environment conditions. While maintaining productivity, precision farming practices help save money and efforts of farmers by empowering them to take informed decisions for applying essential inputs in right quantity at the right time. These practices help mitigate the problems associated with the conventional farming practices such as excessive water wastage, waterlogging, chemical run-off, over-fertilization, etc. [2]. Along with climate change, the depletion of natural resources, the rising global population and the continuous decline in cultivable land pose significant threats to food security [3, 4]. There is an urgent need to address agricultural stressors as illustrated in Figure 12.1.

In 2015, all the UN Member States including both developed and developing nations had agreed upon adopting and effectively implementing the 2030 Agenda of Sustainable Development [5]. Today we as humans are living on the verge of depleting our resources which are vital for sustenance of life on earth including food, drinking water, energy, health and the environment. Thus, globally there is an active demand today to bring some effective solutions on table that help in achieving these goals faster for a better future. The seventeen SDGs need to be addressed by each and every country and there is really some alarming situation if we look into achieving goals like food security, extreme climatic conditions, zero hunger, etc. [6]. For a prosperous future, we need to break our old conventional patterns and bring technology to confront such global challenges as the aim is to achieve

DOI: 10.1201/9781003589730-12

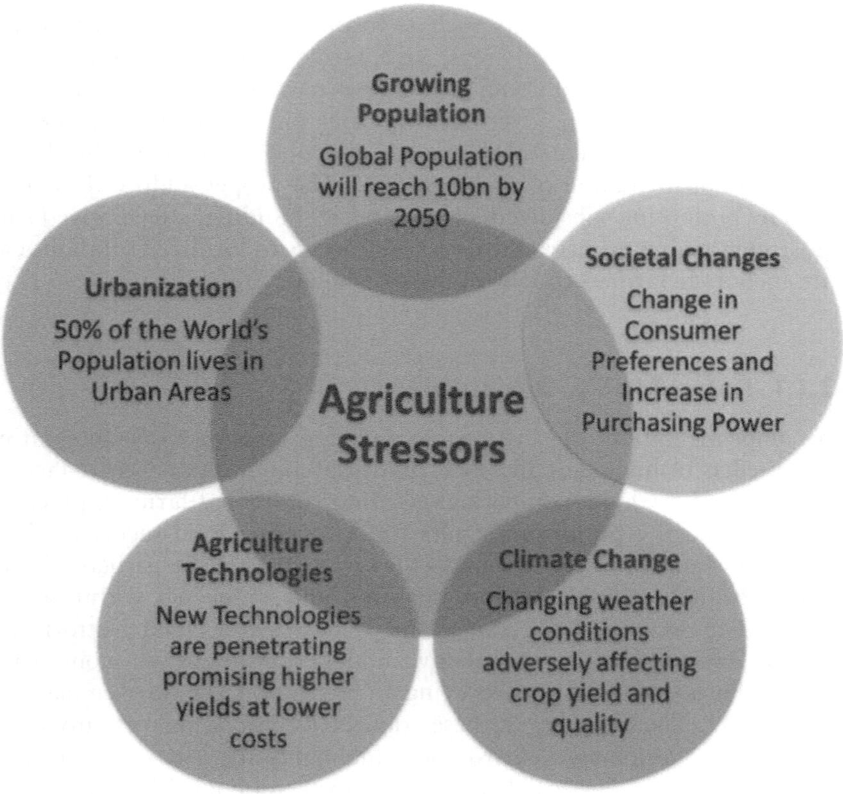

Figure 12.1 Factors contributing to agriculture stressors.

climate-resilient crops. The multi-dimensional, data-driven approach in deciding agriculture frameworks is the need of the hour. The new age of connected farming has already started disrupting the conventional agricultural practices by diminishing the vulnerabilities and limitations associated with old-age patterns. Moreover, the complete transition demands active involvement of government, technocrats, private sector, consumers and other stakeholders to spread proper awareness among the cultivators, educate them, train them and support them which encourage our farmers to adopt such practices [7]. The active role of farmers can be attained by gradually building their trust over such precision farming practices at a larger scale. The farmers must be made aware of such practices, must be encouraged by educating about the long-term benefits and potential outcomes. Farmers in India are reluctant to implement technology-based precision farming practices due to several challenges associated with it. These challenges include high initial investment costs, unawareness, lack of proper knowledge and training, limited exposure to digital tools, fragmented land holdings and less feasibility for marginal farmers [8]. Along with these there are infrastructural constraints such as inconsistent electricity supply and internet connectivity in rural areas. Also, farmers are resistive to change as the traditional mindset of doing farming has passed down through generations [9]. The lacunas in government policies for providing financial support in case of any collateral damage is also one of the major hurdles in effective implementation of such practices. The proper addressal of these issues by promoting subsidized loans, capacity building, cooperative farming, strengthening infrastructure and easy access to localized solutions can only encourage farmers to rely on these technologies.

12.1.1 Role of IoT in agriculture

The implementation of sustainable farming practices by integrating cutting-edge technologies like IoT, Artificial Intelligence (AI), Machine Learning (ML), Cloud Computing, etc. into traditional farming practices helps in achieving promising results for addressing food security. These technologies influence the farming styles by analysing the gathered information through multiple sensors deployed and connected within a network. All the data is being stored and analysed in a cloud platform for fetching better insights into predictive analytics, crop growth monitoring, scheduled process of lighting, spraying and irrigation, data-driven planting patterns, etc. The interplay of these connected sensors and actuators in a self-regulating automated environment highlights the role of IoT in precision agriculture. This IoT-driven system fosters a data-driven approach, enabling a controlled ecosystem that continuously enhances its performance through self-adaptation, self-optimization, self-healing and self-regulation [10]. This level of advancement helps in optimizing resource usage and enhances overall productivity in terms of yield and profits.

For centuries, expansion in the farmland was seen as the only solution to scale up the agricultural production rate. This had adversely affected the environment as the forests were cleared on a large scale, grasslands and wetlands were disturbed and forcibly converted to cultivable lands. The Green Revolution in the mid-20th century further impacted this trend by transforming vast areas into productive farmlands [11]. However, this approach has now reached its limits posing high risks of more deforestation leading to loss of biodiversity, complete extinction of certain species, extreme climate conditions and soil degradation if further expansion continues. As per the studies, globally, now there is a marginal possibility of converting only 5% of the land into cultivable land without severely disturbing the ecological balance [12]. Hence, it is a clear indication that in future there will be no expansion possible in farmlands and to increase the production, existing agricultural practices must be replaced by technology-driven resource-efficient precision farming settings.

Given this challenge, the focus has shifted to sustainable intensification—using technology, precision farming, and better management practices to produce more food on the same amount of land while minimizing environmental impact. By early 2021, technology had advanced enough to boost crop yields from existing farmland through methods like greenhouses, indoor and vertical farming, and precise control of nutrients, water and pest management. Advancements in sensor technologies, the expansion of mobile connectivity, the growth of cloud computing and significant gains in computational power collectively contribute to achieving higher crop yields per hectare [13].

The deployment of an IoT ecosystem in agriculture enables farmers to practice connected farming by remotely monitoring and controlling agricultural processes in real time throughout all stages of the farming cycle. For this a network of different sensors and actuators in conjunction with data acquisition systems, powerful computational platform, connectivity and cloud platform needs to be integrated to take over the conventional farming techniques without overburdening the farmers. Most of the cumbersome processes and tedious tasks will now be taken care of by the fully or semi-automated systems. The complete system will remain active throughout the day and night as per the crop needs and for as long as desired without any human botheration. Farmers will be empowered enough to monitor and control these processes remotely via their phone itself. Any uncommon or undesired symptoms during the crop growth can be easily detected at an earlier stage and farmers will be notified for the same timely.

12.1.2 Benefits of IoT in agriculture

IoT for precision agriculture benefits by providing better economies of scale to big farmers managing their large-scale farms via automation, mitigating the labour costs and hence making operations more cost-effective. IoT is the key technology facilitating the growth of climate-resilient crops to withstand environmental stresses such as drought, extreme temperatures, flooding and pests [14]. These crops play a crucial role in ensuring food security amid climate change. IoT is a data-driven technique which along with other automation techniques helps regulate various factors in order to optimize the crop growth conditions to enhance crop quality and yield with efficient utilization of resources. This advanced farming approach is known as Controlled Environment Agriculture (CEA) [15].

12.2 IoT LAYERED ARCHITECTURE FOR PRECISION FARMING

The reference model of a multi-layered architecture for an IoT-based precision farming system is demonstrated in Figure 12.2.

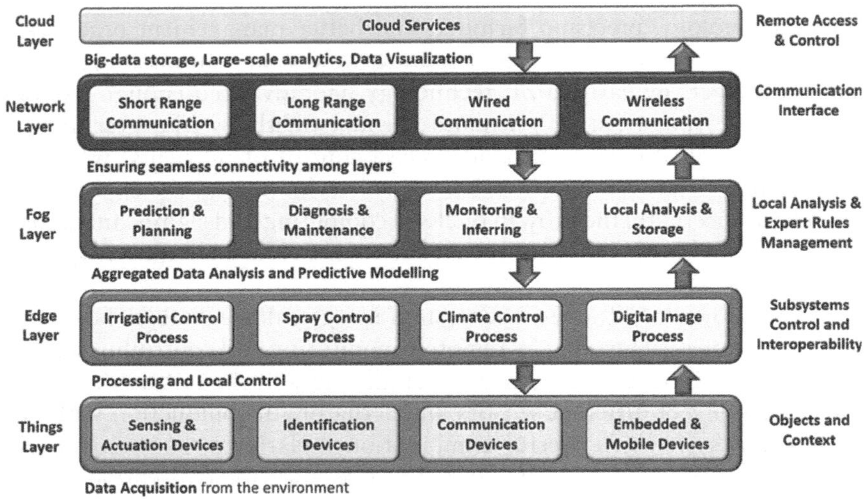

Figure 12.2 Multi-layered IoT architecture for precision agriculture.

The multi-layered stacked arrangement shown in Figure 12.2 integrates diverse technologies to regulate the self-paced precision farming conditions and to optimize other farming processes involved. This data-driven model supports this architecture to enable remote monitoring of parameters in real time and capability to configure and control certain parameters remotely. To enhance the overall yield and bring sustainability in farming, precision farming should be preferred over conventional farming practices. Achieving this is not possible without technological intervention in agriculture practices. The whole system functionality is supported by different layers and each layer works in close conjunction with each other. The seamless connectivity ensures consistency in data flow, data processing and analytics.

1. Things Layer: The things layer is the primary physical layer in the proposed system where diverse sensors, actuators, data-loggers are connected in a mesh network. These devices are connected together to collect data, to analyse it, to make computations, to act upon it and to share it over the internet via a cloud platform. The sensors deployed in the field measure critical parameters related to environment, nutritional value of water, dosing rates, plant growth rate, ambient light intensity, etc. required for a self-paced precision farming set-up. This layer is solely responsible for gathering the actual raw data from the field to extract information and get insights to make predictions and take decisions by observing some patterns. This layer also embraces radiofrequency identification (RFID)-based identifications, precise Global Positioning System (GPS)-based tracking and drone-based remote sensing.

2. Edge Layer: This layer is accountable for collecting the raw data from the physical layer and feed this big data for further processing to the edge devices including high-performance computing platforms and gateways to perform operations like data filtering, data normalization, data preprocessing, etc. Moreover, the switching decisions for multiple actuation processes are also carried out at this layer. The data is compressed before transferring it to the subsequent layer for further processing.
3. Fog Layer: This layer handles advanced data processing using local servers. It performs aggregated data analysis and predictive analytics, enabling real-time decision-making closer to the source. Some processed data is also stored at this layer to reduce cloud dependency. Additionally, this layer enhances security and authentication, acting as a protective gateway between edge devices and the cloud.
4. Network Layer: This layer ensures seamless connectivity between different layers by enabling exchange of data between the sensors, edge devices, fog nodes and cloud platform. This layer establishes wired communication using ethernet, optical fibres; wireless short-range communication between devices via communication interfaces like Bluetooth Low Energy (BLE), Zigbee, Wi-Fi, LoRaWAN; long-range communication via NB-IoT, 4G/5G cellular service and satellite communication. It enables periodic data transfers between IoT devices.
5. Cloud Layer: This is the topmost layer in the hierarchy that facilitates long-term sensor data storage for trend analysis, earlier predictions and optimizations. User can visualize the farm data over the dashboards via mobile/laptop. Farmers can remotely access the farm information and can control some processes remotely from anywhere. Few popular cloud platforms available are AWS IoT Core, Microsoft Azure, Google Cloud.

12.3 IoT-DRIVEN CONTROLLED ENVIRONMENT AGRICULTURAL PRACTICES

A controlled ecosystem providing artificial crop growing conditions free from agronomic constraints by regulating their environment tailored to their needs is known as CEA. The integration of cutting-edge technologies such as IoT, edge computing, AI, ML proved to be the key enablers for the wide adoption of CEA practices globally [16]. IoT systems with high-performance computation, high-precision sensing and seamless connectivity can revolutionize such agriculture practices by enabling real-time monitoring, predictive analytics and automated decision-making. Vertical farming or indoor farming is another referral to CEA and it is the future of farming as it helps in bringing sustainability in agriculture [17]. As CEA follows the data-driven approach, the role of IoT is critical in real-time

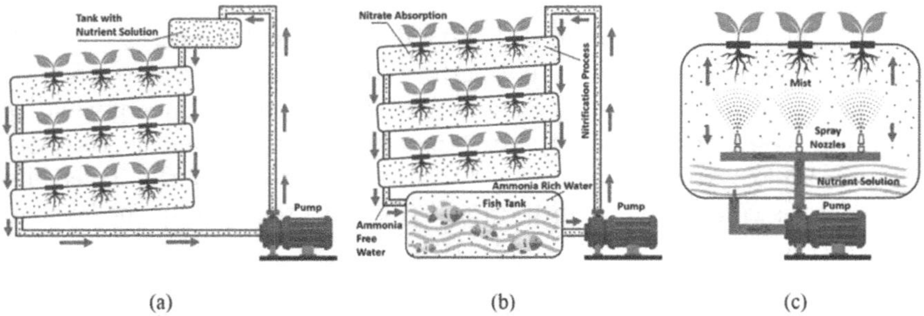

Figure 12.3 Controlled environment agriculture (CEA) under (a) hydroponics set-up, (b) aquaponics set-up, (c) aeroponics set-up.

monitoring, automated climate control, precision irrigation, nutrient management and predictive analytics to optimize crop growth and resource efficiency. Some IoT-based applications are mentioned below:

- CEA involves the deployment of IoT-based, highly responsive climate control systems [18] to regulate environmental factors such as temperature, humidity and light, ensuring optimal conditions tailored to crop-specific requirements.
- CEA encompasses soilless farming techniques such as hydroponics, aeroponics and aquaponics, which are among the top trends in the adoption of precision farming practices. Feedback-driven, self-regulating IoT systems [19] are increasingly being deployed in such CEA settings.
- CEA also incorporates IoT-based artificial lighting via power-efficient light-emitting diodes (LEDs) to mimic natural sunlight, enabling year-round cultivation of crops in vertically arranged stacks [20].
- CEA also covers the mechanism of precision spraying [21].

Although vertical farming is a multi-technology-driven practice and a costly affair till date [22], the integration of IoT-based solutions plays a significant role in making this more viable and feasible for crop cultivation. The use of relevant hardware for sensing, computations and actuation along with networked communication optimizes different operations carried out in vertical farming to make the crop production more competitive and profitable. Figure 12.3 displays three types of vertical farm settings under CEA. These setups are for hydroponics, aquaponics and aeroponics, respectively.

12.4 IoT-ENABLED SMART HYDROPONICS SYSTEM

Hydroponics is a soil-free plant growth mechanism where the plants fulfil their nutritional requirements from submerged roots in a nutrient-rich

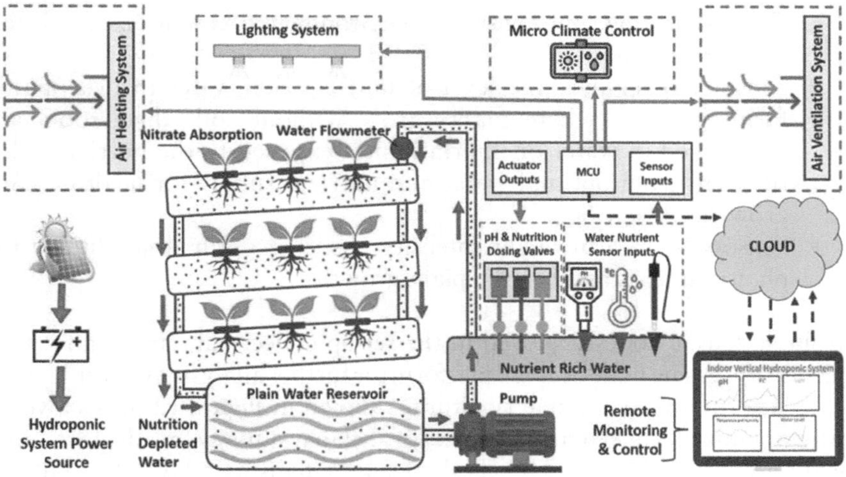

Figure 12.4 A controlled environment agriculture facility for hydroponics.

solution. It's a vertical farming system where plants are sheltered on a tray filled with this nutrient solution. The circulation of this nutrient solution in each tray from the main reservoir, throughout the day via a pump, after fixed intervals, facilitates the recharge process. The hydroponics system integrated with IoT has transformed traditional farming practices, bringing more efficiency, sustainability and intelligence to agriculture [23]. The capability of these systems for real-time monitoring, data-driven decision-making, self-paced multi-process automation and analytics has revolutionized precision farming. In such systems, a multi-sensory environment, the actuators perform based on certain conditions which can be monitored and controlled remotely via a cloud platform, and the analytics can be used for deciding future course of actions [24, 25].

An IoT-based CEA facility for hydroponics is shown in Figure 12.4. The major areas for IoT interplay in a smart hydroponics system include as follows.

12.4.1 Remote monitoring and data acquisition in real time

Multiple sensors are deployed for gathering the information of some vital parameters such as:

- Sensors to gather information about the concentration of nutrients in water by measuring pH levels and electrical conductivity of nutrient-rich water.

- Sensors to measure ambient temperature and humidity levels for precise monitoring and control of environmental conditions to ensure better plant growth.
- Sensors to regulate the level of nutrient-rich water in hydroponic farming trays, where plants grow without soil, their roots are submerged in water or intermittently exposed to the nutrient solution, ensuring optimal water and nutrient supply for healthy plant growth.
- Sensors to measure light intensity ensuring optimal conditions for photosynthesis and healthy plant growth.

Multiple sensor nodes deployed in the set-up collect data and send it to a master processing node or cloud platform for further analysis. A multi-layered IoT platform for precision agriculture is shown in Figure 12.5. It's a vertically organized stacked arrangement of scalable sub-systems deployed in a field for data acquisition. This mesh of sub-systems uses MQTT to publish data to the broker. This collected field data is sent to a cloud platform for more useful insights and analytics.

12.4.2 Dynamic actuation and control

Based on the inputs acquired from sensors, the action takes place by controlling the process of actuation with minimum intervention of humans. For instance:

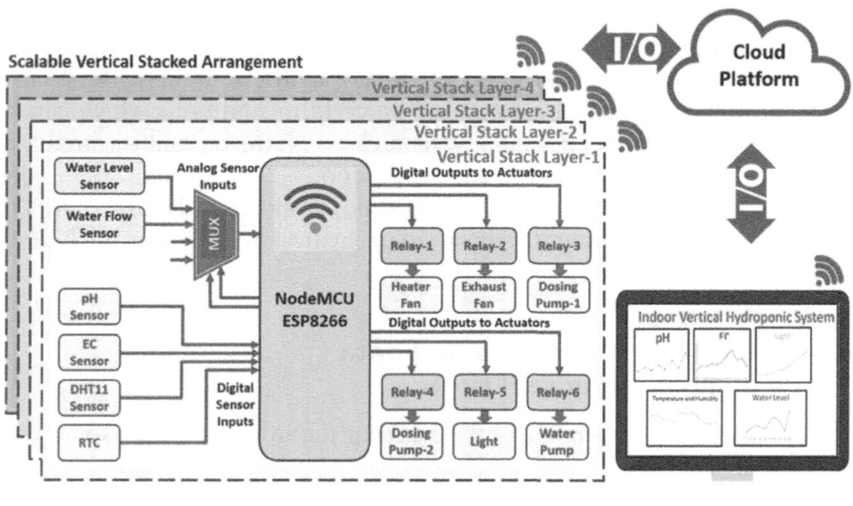

Figure 12.5 Multi-layered IoT platform for precision agriculture.

- In vertical farming, to maintain the appropriate measure of nutrient dosage and water pH levels, the actuators are responsible as they circulate this nutrient-rich water through all the vertical trays [26]. For this the microcontroller takes care of providing proper activation/deactivation commands to the pumps and solenoid valves ensuring consistence and precision in water regulation through the complete vertical set-up.
- Maintaining optimal conditions conducive to create an artificial environment as per the requirement of a specific plant growth, the self-paced switching of appliances is obligatory [27]. This switching is governed by the microcontroller unit (MCU) only as they command multi-channel relays and contactors to control the installations such as air-humidifiers, air heating systems, ventilators, exhaust fans and lights inside the facility.
- Smart irrigation management systems can employ on-off delay timers to regulate water supply with precision, ensuring it is delivered periodically according to the specific needs of each plant. The schematic design of an on-off delay timer is shown in Figure 12.6. A microcontroller-based automatic on-off scheduler is used for controlling the switching of water pump. This system helps in execution of precise irrigation cycles for hydroponic systems which require periodic cycles of irrigation.

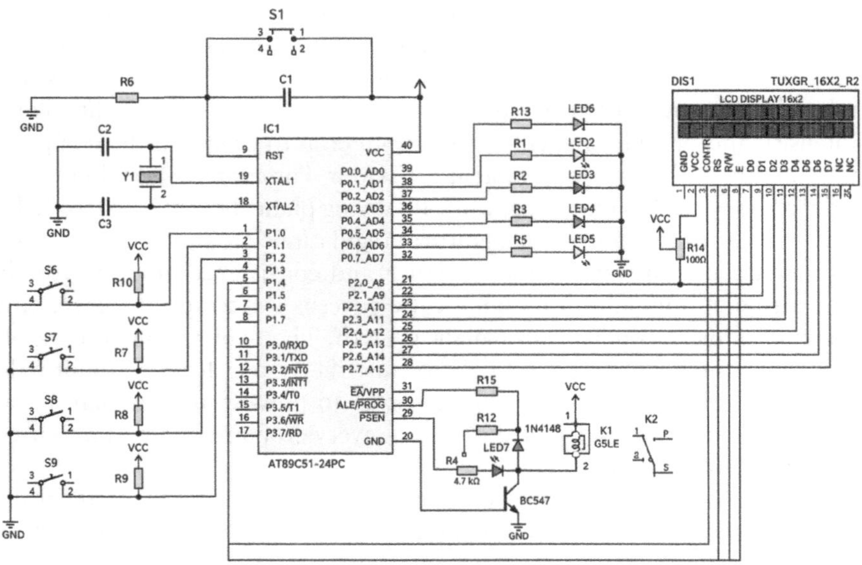

Figure 12.6 Schematic diagram of automatic on-off scheduler for pump control.

12.4.3 Data storage and analytics

The role of IoT in hydroponic farming is to collect the field data continuously through an interconnected multi-sensor network. We may rely on cloud platforms to store and process the gathered information, where it supports detailed analysis to form analytics, visualizations and insights for making predictions. The analytics enable farmers anticipate the plant growth patterns and detect any irregularities at an earlier stage, which is not possible in conventional farming style. The intervention of technology such as IoT in farming practices enhances convenience for farmers to monitor and regulate the crop growth parameters remotely with minimum efforts, more precisely and with better ease. Using the web dashboards or mobile applications, the farmers can review historical data, evaluate past performance, improve farming strategies and maintain optimum environmental conditions to accomplish the goal of improved crop yield.

Detailed Explanation of the IoT-Based Hydroponic Facility: The CEA system illustrated in Figure 12.4 represents a unit-cell for the stacked arrangement of an IoT-based hydroponic set-up specifically designed for vertical farming applications. This system configuration integrates real-time data acquisition systems with high-performance computations, and incorporates smart closed-loop control systems to create optimal environment settings required for plant growth without soil. The proposed framework highlights effectiveness, accuracy and brings sustainability in managing environmental factors.

The Lighting System: In vertical farming, the regulation of sufficient light of suitable wavelength is vital for photosynthesis process supporting plant growth. The exposure to natural sunlight is restricted in such cases and relies on low-power LED-based artificial lighting systems. The light colour, intensity and periodic switching vary from crop to crop and the computational units deployed govern system behaviour. For example, seedlings typically respond well to blue light, while flowering plants benefit from red light.

Air Heating System: Temperature regulation is another important parameter fundamental to the existence and consistence in crop growth. The air-conditioning unit automatically governs the temperature, making the environment conducive for plant growth. The deployment of multiple sensors in the field feeds the MCU to tune the system for any abnormalities and sudden fluctuations in temperature to prevent any thermal stress. The adaptive control safeguard limits for every crop are different and need proper consideration.

Nitrate Absorption: In hydroponic systems, the major plant nutrition depends upon the nitrate compound. A dedicated system must take care of the nitrate concentration in water and flow of nutrient-enriched water in this soilless cultivation system. The flow is controlled through switching of actuators such as solenoid valves governed by MCUs to uniformly distribute this nutrient-enriched water. The system considers continuous feedback obtained from these nitrate sensors to adjust the nutrient concentration levels required for healthy plant growth and improved crop yields.

pH and Nutrition Dosing Valves: In hydroponic systems, it is fundamental to ensure the correct balance of pH and nutrient concentration in water circulated in the set-up. For this the water-quality sensors and automated solenoid valves are usually deployed in the field to keep track of water nutrients and fine-tune the chemical composition of water appropriate for plants. This MCU-based system responds to the feedback received, by regulating the switching cycles of dosing valves with minimum human intervention.

Circulation System for Nutrient-Rich Water: The circulation of this water solution is taken care by pumps regulated by MCU. The operation is governed by microcontroller to monitor the water flow rate, level of pressure, maintaining uniform and efficient distribution throughout. To bring more reliability to the system, a reservoir of plain water is provided as an auxiliary source. To ensure the uninterrupted supply of nutrients to the plants, this microcontroller-based system switches to this reserve automatically to refill the main tank whenever there is a decline in water levels observed.

Monitoring and Control Remotely: One of the major benefits that can be leveraged by deploying such systems is the empowerment one feels by getting remote monitoring and control features. The IoT-enabled system provides remote accessibility to the information gathered from diverse sensors and the real-time visualization of different parameters to the user. User gets the ability to control the switching of actuators remotely. The cloud-based interface is used to collect, store and analyse data acquired from flow-sensors, float-sensors, nutrient sensing probes and the physical parameters of environment. Farmers can access this data-log in real time from any location to observe and get better insights. This enables the farmers to make future predictions and take informed decisions based on the current status. Moreover, the farmers can also configure the system remotely to tune the level of certain parameters such as lighting, temperature, dosing concentration, dosing rate, etc., eliminating the requirement of on-site management. Such ground-breaking features of the IoT-based deployments make it more competent, highly reliable and suitable when the remotely located vertical farming set-up needs to be scaled up with associated challenges and restricted human intervention.

Power Source for Hydroponic System: A reliable power source is an integral part of any hydroponic system as each system has various components like lightening, air heating and pumping, with controlling units relying on electrical power for their proper operation. For safety assurance, the design adds various safeguards against voltage fluctuations and power disruptions. This protects the health and production of crops by maintaining operations like nutrient delivery and regulation.

Management of Nutrient-Depleted Water: The concentration of minerals in soil decreases with plant absorption, and the system would continue to track the nutrient levels using dedicated sensors. The input from the sensors fed to the MCU would allow it to regulate the levels of nutrient by providing appropriate dosage whenever needed. This balances the nutrient levels in the plants.

Plain Water Reservoir: As the water level changes in the tank, the MCU ensures that the pump activates to draw water from the water bodies (reservoir) to the tank for maintaining the proper water level. This in turn prevents the issues that may be caused due to water scarcity.

Water as the Growth Medium: In hydroponic cultivation, instead of soil, water becomes the source of nutrients and oxygen for plant roots. To maintain the appropriate level of the parameters like pH, temperature and nutrient concentration, we use real-time monitoring for automated adjustments to preserve the required proportion of each parameter. This in turn reduces errors and human interaction.

Key Advantages of the System: An IoT integration for smart hydroponic framework allows to control the growth conditions and improvement of crop quality and quantity by reducing human efforts. The system uses real-time analysis and automation which ensures proper usage for water, nutrients and other resources. This benefits farmers as they can remotely manage the operations. In the long term, the system would prove to be cost-effective as it saves expenditure on crop treatments on various scales.

The major advantages of this system include:

1. Automation: Various agricultural functions like pH balance and nutrient management are automatically regulated by the system and can be manually supervised.
2. Precision: By the continuous sensor feedback and control, optimal quantity of various parameters is maintained to ensure proper plant growth.
3. Scalability: Such IoT architecture allows for flexibility in implementation of the system for both small- and large-scale set-up.
4. Remote Accessibility: The system can be monitored and controlled from any remote location as it uses cloud-based interfaces, also for unmanned or distant farms.
5. Sustainability: The system also ensures that resources are utilized to their full potential with eco-friendly practices to align with sustainable development goals.

Real-time operational data transmission to a cloud platform like ThingSpeak is illustrated in Figure 12.7. The data collection from the system can be further analysed to activate specific responses like pumping of water and nutrient levels adjustment. These self-regulating properties of IoT-integrated system play an important role in transforming farming practices.

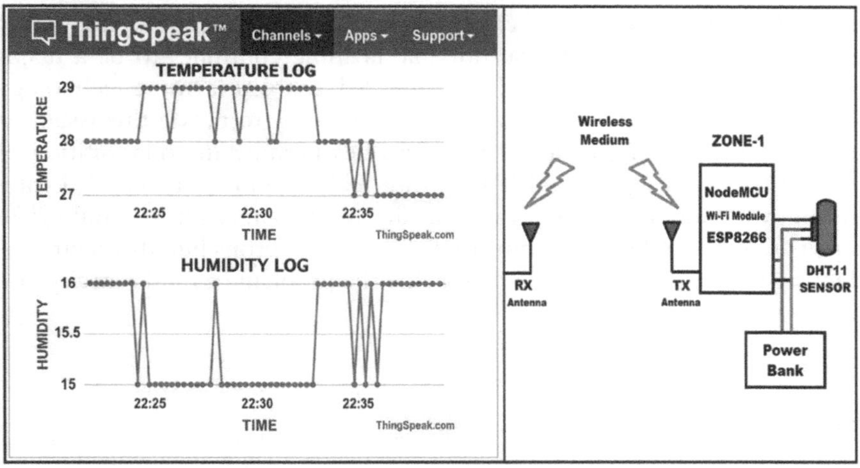

Figure 12.7 Real-time temperature and humidity data-log on ThingSpeak.

12.5 CONCLUSION

The intervention of technology such as IoT in agriculture brought some perceptible transformations in the conventional farming practices, bestowing more sustainable approach to curb the global challenges for food security and climate change. Leveraging the potential of some cutting-edge technologies available including IoT, AI, cloud computing, 5G communication and their convergence empowered the systems to collect data more easily, store large data more reliably, analyse it more rigorously and transmit it more frequently. Such systems aid farmers to manage the farming processes with unprecedented accuracy. The case study of the IoT-based hydroponic system demonstrated the potential of adopting connected approach that redefines the CEA frameworks that include aeroponics, aquaponics and hydroponics. In this chapter, the multi-layered architecture of IoT outlines the role of sensors, actuators, computing platforms to collect data which is further synchronized to the cloud platforms for achieving real-time monitoring, analytics, useful insights for making future predictions, observe patterns by analysing history, autonomous solutions and informed decision-making. This data-led system reduces dependence on human intervention or labour and also empowers farmers to make decisions in a factual manner. The ability to remotely observe and even control the system adds to the scalability, precision factors for all kinds of farm business (small or large). Regardless of its strengths, the adoption of IoT-driven agricultural practices faces challenges like high set-up costs, awareness of technologies and gaps in various infrastructure. To overcome these challenges, collective efforts like collaboration of farmers, researchers, policymakers etc. must be considered. With increasing population on a global scale, the need for sustainable farming continues to be a major step as land for cultivation is decreasing. IoT-enabled farming technology would play a key role in maximizing productivity to tackle this issue. At last, the future of agricultural practices lies broadly in collaboration of various technical aspects of IoT with thorough farming practices. IoT integration with adoption of innovations, education on agriculture would yield results that not only would improve the quality of crops but also their production would increase which can tackle the problem of food scarcity and achieve sustainable results.

REFERENCES

1. J. Wang, Y. Li, and C. Liu, "A Survey of Internet of Things (IoT) in Agriculture," *IEEE Access*, vol. 8, pp. 41528–41545, 2020. DOI: 10.1109/ACCESS.2020.2978180.
2. Muhammad Ayaz, Mohammad Ammad-Uddin, Zubair Sharif, Ali Mansour, and El-Hadi M. Aggoune, "Internet-of-Things (IoT)-Based Smart Agriculture: Toward Making the Fields Talk," *IEEE Access*, vol. 7, Sep 2019. DOI: 10.1109/ACCESS.2019.2932609.
3. Virendra Kumar Sharma, Ashish Kumar Pandey, and Vinay Kumar Singh, "An Intelligent IoT-Based Wireless Sensor Network System for Precision Agricultural Monitoring," *International Journal for Research in Applied Science and Engineering Technology*, vol. 14, no. 3, pp. 2926–2932, March 2026. DOI:10.22214/ijraset.2026.78541.
4. Asad Ul Haq Hashmi, Ghias Uddin Mir, Kashif Sattar, Syed Sajid Ullah, Roobaea Alroobaea, Jawaid Iqbal, and Saddam Hussain, "Effects of IoT Communication Protocols for Precision Agriculture in Outdoor Environments," *IEEE Access*, vol. 12, pp. 46410–46421, March 2024. DOI: 10.1109/ACCESS.2024.3381522.
5. Monica Dutta, Deepali Gupta, Sumegh Tharewal, Deepam Goyal, Jasminder Kaur Sandhu, Manjit Kaur, Ahmad Ali Alzubi, and Jazem Mutared Alanazi, "Internet of Things-Based Smart Precision Farming in Soilless Agriculture: Opportunities and Challenges for Global Food Security," *IEEE Access*, vol. 13, pp. 34238–34268, Feb 2025. DOI: 10.1109/ACCESS.2025.3540317.
6. Dianes David, O. S. Albahri, A. H. Alamoodi, A. S. Albahri, Muhammet Deveci, and Iman Mohamad Sharaf, "Integrating Sensors and Multicriteria Decision Making (MCDM) in Precision Agriculture: A Mini Review," *IEEE Sensors Reviews*, vol. 2, pp. 444–450, Aug 2025. DOI: 10.1109/SR.2025.3596890.
7. Radhika Raina, Kamal Jeet Singh, and Suman Kumar, "Power Efficient and Long Range Precision Agriculture Monitoring System," *IEEE Journal of Radio Frequency Identification*, vol. 9, pp. 330–339, May 2025. DOI: 10.1109/JRFID.2025.3574759.

8. Sarowar Morshed Shawon, Ifaz Ahmed Adan; Md Moniruzzaman, Osama Haramine Sinan, Md Junaed Abed, and Marran Al Qwaid, "Federated Learning Approaches in Precision Agriculture: A Systematic Review of Current Trends, Challenges, and Opportunities," *IEEE Access*, vol. 14, pp. 30513–30538, Feb 2026. DOI: 10.1109/ACCESS.2026.3667548.
9. Kshetrimayum Lochan, Asim Khan; Islam Elsayed, Bhivraj Suthar, Lakmal Seneviratne, and Irfan Hussain, "Advancements in Precision Spraying of Agricultural Robots: A Comprehensive Review," *IEEE Access*, vol. 12, pp. 129447–129483, Aug. 2024. DOI: 10.1109/ACCESS.2024.3450904.
10. Sahira Vasquez, Ciro Allarà, Antonio Altana, Pietro Ibba, Michele Gullino, and Elias Holzknecht, "Modular Impedimetric Sensing Platform for Laboratory and Field Characterization in Precision Agriculture," *IEEE Transactions on AgriFood Electronics*, vol. 3, issue 2, pp. 328–335, Aug. 2025. DOI: 10.1109/TAFE.2025.3591564.
11. A. A. A. Al-Turjman and S. Zeadally, "Toward Smart Farming: Systems Prototype Using IoT," *IEEE Access*, vol. 5, pp. 5998–6009, 2017.
12. S. Sundaram and K. T. Shanmugam, "IoT Based Smart Agriculture Monitoring System," *2018 International Conference on Smart Electronics and Communication (ICOSEC)*, Tamil Nadu, India, 2018, pp. 1–4.
13. K. S. Balamurugan, and A. Sivakami, "LoRa-IoT Based Self-Powered Multi-Sensors Wireless Network for Next Generation Integrated Farming," *International Journal of Scientific & Technology Research*, vol. 8, issue 10, pp. 1527–1533, Oct. 2019.
14. Muhammad Luthfi Lokman, Kah Yoong Chan, Yik Tian Ting, Chu Liang Lee, Gwo Chin Chung, and Wai Leong Pang, "LoRa Based IoT Enabled Sensor Networks for Plantations," *Journal of Engineering Technology and Applied Physics*, vol. 6, no. 1, pp. 16–24, March 2024. DOI: https://doi.org/10.33093/jetap.2024.6.1.3.
15. Sathiyanathan S, Jagadeesh D, Kartheeban K, Manoj C S, and Nandhakumar M, "An IoT-Driven Smart Agriculture System for Intelligent Crop Disease Detection, Health Monitoring, and Precision Farming Applications," *5th International Conference on Communication, Computing and Electronics Systems (ICCCES)*, Jan. 2026. DOI: 10.1109/ICCCES62661.2026.11436788.
16. Ahmed Khattab, Ahmed Abdelgawad, and Kumar Yelmarthi, "Design and implementation of a cloud-based IoT scheme for precision agriculture," *28th International Conference on Microelectronics (ICM)*, Feb. 2017. DOI: 10.1109/ICM.2016.7847850.
17. Monica Dutta, Deepali Gupta, Sumegh Tharewal, Deepam Goyal, Jasminder Kaur Sandhu, Manjit Kaur, Ahmad Ali Alzubi, and Jazem Mutared Alanazi, "Internet of Things-Based Smart Precision Farming in Soilless Agriculture: Opportunities and Challenges for Global Food Security," *IEEE Access*, vol. 13, pp. 34238–34268, Feb. 2025. DOI: 10.1109/ACCESS.2025.354031.
18. A.-L. Balogun, N. Adebisi, I. R. Abubakar, U. L. Dano, and A. Tella, "Digitalization for transformative urbanization, climate change adaptation, and sustainable farming in Africa: Trend, opportunities, and challenges," *Journal of Integrative Environmental Science*, vol. 19, no. 1, pp. 17–37, Dec. 2022.

19. Olakunle Elijah, Tharek Abdul Rahman, Igbafe Orikumhi, Chee Yen Leow, and MHD Nour Hindia, "An Overview of Internet of Things (IoT) and Data Analytics in Agriculture: Benefits and Challenges," *IEEE Internet of Things Journal*, vol. 5, issue 5, pp. 3758–3773, Oct. 2018. DOI: 10.1109/JIOT.2018.2844296.
20. C. Brewster, I. Roussaki, N. Kalatzis, K. Doolin, and K. Ellis, "IoT in Agriculture: Designing a Europe-wide Large-scale Pilot," *IEEE Communications Magazine*, vol. 55, no. 9, pp. 26–33, Sep. 2017.
21. S. Li, H. Zhang, and Y. Jiang, "IoT-Based Smart Irrigation System with Self-Paced Feedback Control," *IEEE Transactions on Industrial Informatics*, vol. 69, no. 3, pp. 412–425, 2023.
22. Q. Chen, L. Wang, and Y. Zhang, "Adaptive Irrigation Control System for Precision Agriculture Using IoT Technology," *Computers and Electronics in Agriculture*, vol. 138, pp. 215–226, 2023.
23. N. Dlodlo and J. Kalezhi, "The Internet of Things in agriculture for sustainable rural development," *International Conference on Emerging Trends in Networks and Computer Communications (ETNCC)*, pp. 13–18, May 2015. DOI: 10.1109/ETNCC.2015.7184801.
24. S. Ivanov, K. Bhargava, and W. Donnelly, "Precision farming: Sensor analytics," *IEEE Intelligent Systems*, vol. 30, no. 4, pp. 76–80, July 2015. DOI: 10.1109/MIS.2015.67.
25. E. F. I. Raj, M. Appadurai, and K. Athiappan, "Precision farming in modern agriculture," *Smart Agriculture Automation Using Advanced Technologies: Data Analytics and Machine Learning, Cloud Architecture, Automation and IoT.* Berlin, Germany: Springer, pp. 61–87, Jan. 2022.
26. O. Elijah, I. Orikumhi, T. A. Rahman, S. A. Babale, and S. I. Orakwue, "Enabling smart agriculture in Nigeria: Application of IoT and data analytics," *IEEE 3rd International Conference on Electro-Technology for National Development (NIGERCON)*, Owerri, Nigeria, pp. 762–766, Nov. 2017. DOI: 10.1109/NIGERCON.2017.8281944.
27. S. Wolfert, L. Ge, C. Verdouw, and M.-J. Bogaardt, "Big data in smart farming—A review," *Agricultural Systems*, vol. 153, pp. 69–80, May 2017.

Index

Note: **Bold** page numbers refer to tables and *italic* page numbers refer to figures.